Direct Current Circuit Analysis

Robert G. Seippel
Instructor, Electronics
College of the Canyons,
Valencia, California

Roger Lincoln Nelson
Development Engineer
Honeywell Corporation
Minneapolis, Minnesota

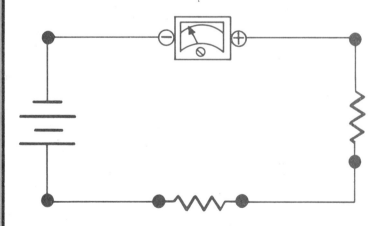

 American Technical Society ▪ Chicago 60637

COPYRIGHT © 1975
BY AMERICAN TECHNICAL SOCIETY

All Rights Reserved

Library of Congress Catalog Number: 74-24889
ISBN 0-8269-1587-6

No portion of this publication may be reproduced by any process such as photocopying, recording, storage in a retrieval system or transmitted by any means without permission of the publisher.

123456789 75 987654321

Printed in the United States of America

Preface

The analysis of direct current (dc) circuits is almost an ancient art. Almost everyone from Benjamin Franklin to Penelope Priscilla Pendergast has written a dissertation on the analysis of these circuits.

In most cases these texts are thorough, with circuit coverage complete, but they are not always brief and simple. In some cases circuit coverage is a reflection of the writer's background and the compilation of a wealth of technical knowledge on the subject.

Great detail is understandably necessary at times, depending on the particular circumstance. There is a definite need and place for detail and unavoidable complexity, especially in research and science, and in most engineering. No writer would attempt to compile any technical material without valid reference material, which is supplied by the tireless efforts of reliable writers.

Regardless of the thousands of papers written on the subject, however, instructors still find themselves searching for a simpler method for students to analyze dc circuits. It is the objective of this book to provide a simple mode for doing this. Procedures herein have been constructed in cookbook style so they may be followed closely by the beginner, while providing a quick reference to the journeyman.

Artwork is drawn in stages, to establish continuity. Some electrical basics are also covered, but in these cases details are kept to "nutshell" descriptions.

The writer acknowledges all others who have attempted an easier way and recognizes that this book could not possibly fill all the needs of everyone involved in the details of dc circuit analysis.

Hopefully, the simplified presentation offered here can aid toward a better understanding of the most used laws and theories on the subject without requiring the reader to wade through foggy swamps of complexity.

THANKS TO: Robert G. Seippel

Hazel, my wife—Typist
Dal Fitts—Technical Editor

TABLE OF CONTENTS

Section	Page No.
1 **BASIC FACTS ABOUT DIRECT CURRENT**	1–14
What is Direct Current?	2
Resistance in DC Circuits	4–8
Insulators	8
Ohm's Law	8–11
Direct Current Power Formulas	11–14
2 **SERIES DC CIRCUIT**	15–20
Series DC Circuit Formulas	15–17
Voltage Division	17–19
"Opens" in a Series Circuit	19–20
"Shorts" in a Series Circuit	20
3 **PARALLEL DC CIRCUIT**	21–27
Parallel DC Circuit Formulas	21–25
"Opens" in a Parallel Circuit	26
"Shorts" in a Parallel Circuit	26–27
4 **THE WHEATSTONE BRIDGE**	28–29
5 **SERIES PARALLEL DC CIRCUIT**	30–36
Voltage in a Series Parallel DC Circuit	30–31
Resistance in a Series Parallel DC Circuit	31–32
Current in a Series Parallel DC Circuit	32–33
Voltage Division in a Complex Circuit	33–34
Calculating Total Resistance in a Complex Circuit.	34–36
6 **KIRCHOFF'S LAWS FOR CURRENT AND VOLTAGE** ...	37–39
Kirchoff's Current Law	37–38
Kirchoff's Voltage Law	38–39
7 **MESH CIRCUITS**	40–42
Mesh Circuit Analysis	41–42
8 **SUPERPOSITION PRINCIPLE**	43–45
Method of Calculation by Superposition Principle .	43–45
9 **THEVENIN'S THEOREM**	46–48
Method of Calculation Using Thévenin's Theorem.	46–48

10	NORTON'S THEOREM..........................	49–51
	Method of Calculation Using Norton's Theorem..	49–51
11	EQUIVALENT CIRCUITS.......................	52–54
	Method of Calculating an Equivalent Circuit....	52–54
12	MILLMAN'S THEOREM........................	55–58
	Method of Calculation Using Millman's Theorem.	55–58
13	WYE AND DELTA NETWORKS..................	59–63
	Wye to Delta Resistance Conversion..........	60–61
	Delta to Wye Resistance Conversion..........	61–62
	Bridge Network Resistance Solution..........	63
14	NODAL METHOD OF LOOP VOLTAGES...........	64–67
	Method of Calculation Using Nodal Analysis....	64–67
15	RECIPROCITY THEOREM......................	68–70
	Method of Calculation Using Reciprocity Theorem	68–70
16	COMPENSATION THEOREM...................	71–73
	Method of Calculation Using Compensation Theorem..	71–73
17	THE BATTERY................................	74–76
	Batteries Connected in Series................	75–76
	Batteries Connected in Parallel..............	76
18	ELECTRICAL MEASUREMENTS FOR DC CIRCUITS...	77–80
	How to Measure Current in a DC Circuit.......	77
	How to Measure Voltage in a DC Circuit.......	78
	How to Measure Resistance.................	78–79
	How to Measure Continuity.................	79–80

APPENDIX A: **ELECTRICAL/ELECTRONIC SAFETY**...... 81–85

APPENDIX B: **CALCULATING WITH POWERS OF TEN**.... 86–99

APPENDIX C: **OHM'S LAW AND POWER WHEEL**....... 100

APPENDIX D: **DIRECT CURRENT FORMULAS**......... 101–102

APPENDIX E: **SQUARES, CUBES, ROOTS, AND RECIPROCALS OF NUMBERS FROM 1 TO 100**....................... 103–105

LIST OF ILLUSTRATIONS

Fig. No.   *Page No.*

1. Fixed Resistors 5
2. Schematic Symbols for Resistance 5
3. Variable Resistors 6
4. Resistor Color Coding 7
5. Ohm's Law Triangle 9
6. Series DC Circuit Analysis 15
7. Voltage Division 18
8. Voltage Division Used for Transistor Biasing 18
9. Open in Series Circuit 19
10. Short in Series Circuits 20
11. Parallel DC Circuit Analysis 22
12. Resistance Totals in a Parallel Circuit with 3 or more Resistors 24
13. Open in a Parallel Circuit 26
14. Short in a Parallel Circuit 27
15. The Wheatstone Bridge 28
16. Total Voltage Calculations in a Series Parallel DC Circuit 31
17. Total Resistance Calculations in a Series Parallel DC Circuit 32
18. Total Current Calculations in a Series Parallel DC Circuit 33
19. Voltage Division in a Complex Circuit 34
20. Equivalent Circuit Method for Calculating Total Resistance in a Complex Circuit 35
21. Kirchoff's Current Law 37
22. Kirchoff's Current Law in a Complex Circuit 38
23. Kirchoff's Voltage Law 39
24. Mesh Circuit Analysis 40
25. Superposition Principle 44
26. Thévenin's Theorem 47
27. Norton's Theorem 50
28. Conversion between Thévenin and Norton 53
29. Conversion between Norton and Thévenin 53
30. Millman's Theorem 56
31. Pi, Delta, Tee, and Wye Networks 59
32. Wye to Delta and/or Delta to Wye Circuit Conversions 60
33. Bridge Network Resistance Solution 62
34. Nodal Method of Loop Voltage Analysis 65
35. Reciprocity Electronic Network 69

36. Compensation Theorem 72
37. Batteries Connected in Series 75
38. Batteries Connected in Parallel 76
39. Measuring Current in a DC Circuit 77
40. Measuring Voltage in a DC Circuit 78
41. Measuring Resistance 79
42. Testing for Continuity 79

LIST OF TABLES

Table 1. Battery Cell Electrodes and Their Electrolyte 74
Table B-1. Standard Prefixes, Their Symbols
          and Magnitudes 88

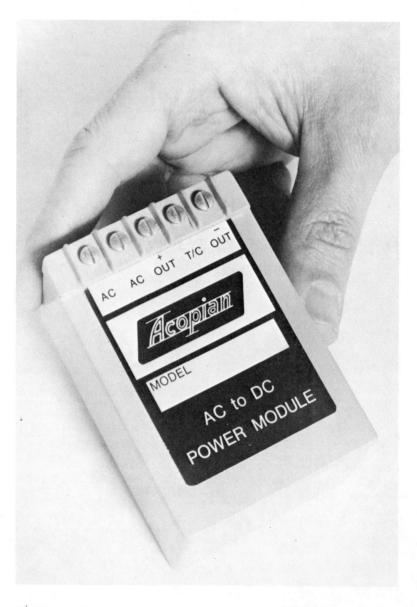

Miniature, regulated power supply changes 60 hertz, 120 vac input to non-pulsating, direct current for use in electronic circuits. (Courtesy of Acopian Corp., Easton, Pa.)

Section 1

Basic Facts About Direct Current

The common characteristics of all types of electrical currents are their variations in amplitude, time, and direction. Direct current was the first current to be widely used and understood. Its applications are relatively simple. However, it has some serious limitations, among which are the facts that it does not develop waveforms and cannot be transmitted (radiated) by an antenna.

If direct current (dc) as we know it today is transmitted by wire for any great distance, the power loss because of wire resistance would tend to be nearly as great as the power originally transmitted. Therefore direct current is impractical to transmit except for a very short distance. It is therefore most useful for fixed-position jobs such as in power machinery, driving mechanisms, and portable machinery that has no immediate access to a supply of alternating current. Direct current is also used extensively for relay controls and solenoid switching.

Probably the best use of dc electrical power is for electronic circuitry, where the direct current provides a "platform" for the ac signals to ride upon or develop around. All electronic circuitry has some amplitude of dc with which the circuit is biased. This level of dc is increased or decreased by varying ac or dc signals.

Direct Current Circuit Analysis

The purety of dc power is always questionable. No absolutely pure, smooth direct current source is available; it all varies somewhat. Designers go to great detail to make dc power supplies that do not vary in amplitude, to ensure that signal development around the dc levels are not affected by changing dc levels. To a practical extent they succeed. Perhaps, then, the greatest advantages of dc power are its characteristics of constancy in amplitude, time, and direction.

WHAT IS DIRECT CURRENT (DC)?

Direct current is electrical current that flows in one direction. This sounds basic. However, the theories concerning the flow of current are numerous and extensive, and some are far from basic. With study, though, all theories fall into place. The scope of this book does not allow space for a detailed discussion of atomic structures, but since the book is directed toward direct current analysis, we feel that it is necessary to say just a few words about the nature of electrical currents. In basic physics, the study of atoms reveals that electrons are in orbit around a nucleus consisting of protons and neutrons. These three types of atomic particles are basic to each atom, and each kind of atom is made up of a different number of orbiting electrons which make it unique.

In the neutral state, each atom has the exact number of electrons in orbit that it takes to make the atom electrically neutral. For instance, a copper atom has 29 electrons (negative charges) in orbit and 29 protons (positive charges) in its nucleus. The outside ring or shell of the atom has loosely held electrons that may be easily removed from their orbit. When removed, the electron becomes a free electron. The atom is now positively charged and is called an *ion*.

Good Conductors of Electricity

A good conductor of electricity is a material that has a large number of free electrons. Gold, silver, and copper fall in this category. Silver is a better conductor than copper, but economic reasons make copper the prevalent wire material used today. Copper has much greater tensile strength than silver, and is also more easily soldered. Aluminum is probably the second most popular metal used for electrical conductors. It is especially useful because of its relatively low cost and its light weight (approximately ⅓ that of copper).

Electron Flow

Electron flow through a conductor is by two methods. First, the free electrons move from one atom's orbit to another. Second, free electrons flow at random throughout the wire without ever going into an atomic orbit. The free electrons are assumed to be evenly distributed throughout the wire or conductor.

If the conductor is open at either end the loop is said to be an *open circuit*. There is no unidirectional electron flow in an open circuit because most electrons are in a state of random motion. When electrical pressure (voltage) is applied by means of an electron pump (battery or generator) connected between the ends of the conductor, the free electrons are forced to flow around the loop. The loop is now said to be a *closed circuit*.

The more pressure (voltage) is applied to this closed circuit, the more electrons break away from their orbits and flow in spaces between the atoms. If the pressure (voltage) is applied constantly at the same strength, the current or electron flow is also constant and is called *direct current*.

If the pressure (voltage) varies in strength (amplitude), the current is said to be *pulsating direct current*. Pulsating current is associated with alternating current and is not intended to be covered in this book. If the pressure (voltage) varies both in amplitude and direction, the current or electron flow through the conductor is called *alternating current*. In all cases electrical current flow is measured in amperes, the amount of electrons that pass a given point per second is estimated as 6.28×10^{18} per ampere. Obviously, it is easier to talk about amperes than electrons.

Conventional Versus Electron Current Flow

"Conventional current" flow is sometimes misconstrued with electron current flow. Electron flow is the flow of free electrons migrating from atom to atom or through atom interspaces. Electron flow is then justifiably called current flow because these electrons move en masse from the "negative" pole of a power supply to the "positive" pole through a conductor.

Conventional current flow was theorized by Benjamin Franklin, who said that current flow was from the positive terminal of a battery to the negative terminal. Franklin was talking about the *internal* current flow in a battery and not the *external* circuit. Internal to a battery there are positively and negatively charged atoms called *ions*. The negatively charged ions at the negative battery terminal give their excess electrons to the external circuit at the negative terminal. The positively charged ions migrate to the positive terminal of the battery and attract free electrons from the external circuit.

The reader will note that the excess electrons do not move through the battery but are carried by ions. Thereby the electrons appear to flow from the positive terminal to the negative terminal.

In the external circuit, however, current flows only in the direction that electrons move. That direction is always from negative to positive.

RESISTANCE IN DC CIRCUITS

The primary electronic component in direct current circuits is resistance. This resistance may come in many forms such as resistors, light bulbs, motors, heaters, solenoids, etc. To keep from complicating our elementary look at dc circuitry, we shall limit our circuit analysis discussion to resistors.

Resistors are either composition or wire-wound. Composition resistors are usually made from carbon and clay, while wire-wound resistors are made from wire. Composition resistors are used for high resistance and wire wound for low resistance. With low resistance, current and power increase; therefore current and power ratings for wire-wound resistors are large.

Each resistor has a built-in tolerance. For instance, you may purchase a 1000 ohm resistance with a 20% tolerance. This would be a guarantee that the resistor will range in resistance somewhere between 800 and 1200 ohms. On the other hand, you may purchase a 1000 ohm resistor with a 10% tolerance. This would be a guarantee that the resistor will range in resistance somewhere between 900 and 1100 ohms. Resistors with 1% or better tolerance are called *precision resistors*.

Each resistor also has a temperature coefficient which tells us, basically, that the resistor changes in ohmic value as temperature changes. A positive temperature change for a resistor with a positive coefficient would note an increase in resistance with a temperature increase. The reverse is true for a decrease in temperature. A positive temperature change for a resistor with a negative coefficient would note a decrease in resistance with a temperature increase. The reverse is then true for decrease in temperature. Metals, from which some resistors are manufactured, have positive temperature coefficients. Carbon resistors have a negative temperature coefficient. Each resistor has a manufacturer's power rating.

Resistors are built to withstand a specified amount of power without being destroyed. For instance, standard carbon type resistors are built with power ratings of $1/8$, $1/4$, $1/2$ and 1 watt. In application, these resistors would be used in circuits that are subject to 50% of their power ratings. Resistors are constructed with either a fixed resistance or a variable resistance. Resistor types are discussed in the next several paragraphs.

Fixed Resistors (See Fig. 1.)

Fixed resistors are constructed to have a fixed value of resistance with a set tolerance. They are made of wire, metal, or a composition such as carbon. They attach by lugs or have wire-soldered ends. In

Basic Facts about Direct Current

Fig. 1. Fixed Resistors

Fig. 1 are several types of fixed resistors. Each fixed resistor is recognized on a schematic by the same symbol. See Fig. 2 for schematic symbols. Some of the various types of fixed resistors are as follows:

Carbon and Clay Resistors. These are the standard resistors used most commonly in electronic equipment. They are made of a composition of carbon and clay and can be used in many applications. The carbon resistor has a negative temperature coefficient. That is, when temperature increases the carbon resistor's ohmic value decreases, and vice-versa. The carbon resistor comes in values of $1/8$, $1/4$, $1/2$ and 1 watt packages and usually has tolerance values of 20% 10% or 5%.

Wire-Wound Resistors. Wire-wound resistors are made with wire wrapped around an insulator such as glass or ceramic. They are usually able to withstand rugged duty because they are large and have a hard coating of ceramic around them. Ohmic values of wire-would resistors are usually within 1% tolerance. Therefore they are called precision resistors. The disadvantage of wire-wound resistors is that they are expensive to make and may effect inductive reactance in high-frequency circuits because of the wire coils.

Metal-Film Resistors. Metal-film resistors are made by heating metal alloys, then vaporizing them on a film or a cylinder of glass,

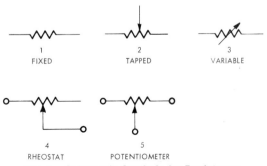

Fig. 2. Schematic Symbols for Resistance

5

then hermetically sealing with glass. Such resistors are manufactured with an ohmic tolerance such as 0.1% and a zero temperature coefficient. Not surprisingly, these resistors are very expensive.

Tapped Resistors. Tapped resistors are variable slide resistors that are fixed by the manufacturer from a slide or by other variable techniques. They are not normally changed from this position and often have a ceramic covering.

Variable Resistors. (See Fig. 3.)

Variable resistors are resistors that can change their ohmic value by some mechanical method. They come in various forms and have tolerances and power ratings just as fixed resistors do. See Fig. 2 for schematic symbols. Some of these variable resistors are as follows:

Slide Resistor. The slide resistor contains a metal sleeve which is wrapped around a wire-wound resistor so that a specific value of resistance may be picked off. The slide may be fixed at any position along the wire-wound resistor and may be adjusted at random.

Potentiometer. The potentiometer is a circular fixed resistor with a rotating contact that picks off a specific resistance. A potentiometer has three terminals and may be of any resistance value such as 0 to 1K ohm or 0 to 10 megohm. The potentiometer may have a single turn or many turns. Many-turn potentiometers are called *helipots*. Potentiometers may rotate through turns up to 360 degrees. Therefore a half turn would provide a resistance value of half the total resistance.

Rheostat. The rheostat is recognized in a circuit as having two terminals. It is built in the same manner as a potentiometer but is usually capable of handling higher current and is used to vary the amount of current in a circuit.

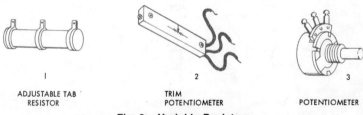

1
ADJUSTABLE TAB RESISTOR

2
TRIM POTENTIOMETER

3
POTENTIOMETER

Fig. 3. Variable Resistors

Basic Facts about Direct Current

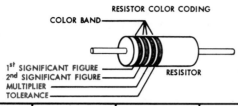

RESISTOR COLOR CODING

BAND COLOR	1ST SIGNIFICANT FIGURE	2ND SIGNIFICANT FIGURE	MULTIPLIER	TOLERANCE
BLACK	0	0	1	---
BROWN	1	1	10	1%
RED	2	2	100	2%
ORANGE	3	3	1,000	3%
YELLOW	4	4	10,000	4%
GREEN	5	5	100,000	---
BLUE	6	6	1,000,000	---
VIOLET	7	7	10,000,000	---
GRAY	8	8	100,000,000	---
WHITE	9	9	1,000,000,000	---
GOLD			0.1	5%
SILVER			0.01	10%
NO COLOR				20%

EXAMPLE:

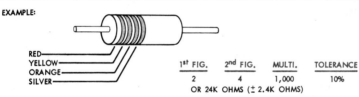

Fig. 4. Resistor Color Coding

Resistor Color Coding (See Fig. 4.)

Fig. 4 illustrates resistor coding. You will note that colored bands on the resistor begin near one end. Reading from this end, the first band represents the 1st significant figure, the second band the 2nd significant figure, and the third band the multiplier. In the example shown, the red band is first, representing the number 2. The second band is yellow, representing the number 4. The third band is orange, representing the multiplier 1000. Therefore the value of this particular resistor is 24,000 ohms. The silver band indicates the tolerance of 10%. This tolerance would provide the user with a resistor value of 24,000 (±2,400) ohms. If there is no fourth band the tolerance on the resistor is 20%.

Resistor Values Below 10 Ohms

Some resistors have values less than 10 ohms. These resistors have a third band which is gold or silver. The gold band instructs to multiply the first two digits by 0.1. Therefore red (2), yellow (4), and gold (multiply by 0.1) bands on a resistor would represent a resistor with a value of 2.4 ohms. If this same resistor had a silver band in place of the gold band, the first 2 band values would be multiplied by 0.01. The value of this resistor would then be 0.24 ohms.

Standard Sizes of Resistance

It is not practical to manufacture and store all different sizes of resistors. Therefore standard values are generally used throughout the industry. These standards are readily recognized by electrical and electronic technicians. For instance, the 4.7 ohm, 47 ohm, 470 ohm, 4.7 k ohm, and 4.7 megohm resistors are standards. Also the 2.2 ohm, 22 ohm, 220 ohm, 2.2 K ohm, 22 K ohm, 220 K ohm and 2.2 megohm resistors are standards.

Resistor supply vendors' catalogs should be consulted for other standard size resistors.

The Thermistor

The thermistor is a device constructed to change resistance value as temperature changes. The word *thermistor* combines the words *thermal* and *resistance*. In an actual circuit, if the resistance of the total circuit increases in value, the thermistor resistance would decrease in value, thereby balancing the total resistance level of the circuit.

INSULATORS

An insulator is a material that has very large resistance. That is, in the millions of ohms and larger. Since insulators have large resistance, very little current flows through them. This makes them good devices to isolate conductors of electricity and therefore prevent cross-current flow.

Insulators are made of many materials. Some materials used are ceramic, paper, glass, oil, mica, porcelain, plastic and rubber.

Another purpose of the insulator is as a *dielectric*. Insulators make possible the storage of an electric charge; therefore they are used as the dielectric material between the two charge plates in a capacitor. This is important chiefly in the design of ac circuits. Other common uses of insulators are for terminal strips and wire coverings.

OHM'S LAW (See Fig. 5.)

Voltage, current, and resistance relationships are defined by Ohm's law. This law states that current in a direct current electrical

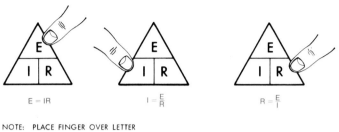

NOTE: PLACE FINGER OVER LETTER OF UNKNOWN, THEN KNOWN VALUES AND THEIR RELATIONSHIP WILL BE UNCOVERED.

Fig. 5. Ohm's Law Triangle

circuit is proportional to voltage and inversely proportional to resistance. The following formulas are used to calculate one of these values if the other two are known.

① $E = IR$ *WHERE:* E = voltage

② $I = \dfrac{E}{R}$ I = current flow

③ $R = \dfrac{E}{I}$ R = resistance

In Fig. 5, a triangle is used to demonstrate these three formulas. Placing your finger on the unknown value will give the formula for that value. For instance, if you place your finger over *I*, then *E* over *R* remains. This indicates that *E* should be divided by *R* to get the value *I*. Place the finger over *E*, and *I* beside *R* remains. This indicates that *I* should be multiplied by *R* to get the value *E*. Place the finger over *R*, then *E* over *I* remains. This indicates that *E* should be divided by *I* to get the value *R*. This procedure will work on all directly proportional equations such as this. Formula variations follow.

Formula Variations of Ohm's Law

BASE FORMULA: $E = IR$

LEGEND:
E = Voltage in Volts
I = Current in Amperes
R = Resistance in Ohms

Direct Current Circuit Analysis

ANALYSIS:

① *If "E" is unknown*

GIVEN:
I = 2 amps
R = 4 ohms

E = IR
E = 2 × 4
E = 8 volts

② *If "I" is unknown*

GIVEN:

E = 8 volts

R = 4 ohms

$I = \dfrac{E}{R}$

$I = \dfrac{8}{4}$

I = 2 amps

③ *If "R" is unknown*

GIVEN:

E = 8 volts

I = 2 amps

$R = \dfrac{E}{I}$

$R = \dfrac{8}{2}$

R = 4 ohms

Varying Values Using Ohm's Law (For Purposes of Calculation)

① *If voltage increases* while resistance stays the same, current will increase and vice versa.

EXAMPLE:	INCREASE IN E	DECREASE IN E
E = 10 volts	E = 15 volts	E = 5 volts
I = 2 amperes	I = 3 amperes	I = 1 ampere
R = 5 ohms	R = 5 ohms	R = 5 ohms

② *If current increases* while resistance stays the same, voltage will also increase and vice versa.

EXAMPLE:	INCREASE IN I	DECREASE IN I
E = 10 volts	E = 15 volts	E = 5 volts
I = 2 amperes	I = 3 amperes	I = 1 ampere
R = 5 ohms	R = 5 ohms	R = 5 ohms

③ *If resistance increases* while current stays the same, voltage will also increase and vice versa.

EXAMPLE:	INCREASE IN R	DECREASE IN R
E = 10 volts	E = 20 volts	E = 4 volts
I = 2 amperes	I = 2 amperes	I = 2 amperes
R = 5 ohms	R = 10 ohms	R = 2 ohms

④ *If resistance increases* while voltage stays the same, current will decrease and vice versa.

EXAMPLE:	INCREASE IN R	DECREASE IN R
E = 10 volts	E = 10 volts	E = 10 volts
I = 2 amperes	I = 1 ampere	I = 5 amperes
R = 5 ohms	R = 10 ohms	R = 2 ohms

Conductance

Each conductor of electricity has resistance and conductance. Resistance, as you recall, is the opposition to current flow. Conductance, however, relates to a conductor's ability to allow current to flow. In mathematical terms, conductance (G) is the reciprocal of resistance and can be calculated using the formula $G = \frac{1}{R}$. Conductances in parallel are totaled by use of the formula $G_T = G_1 + G_2$ which is the reciprocal of the parallel resistance total calculation $\frac{1}{R_T} = \frac{1}{R_1} + \frac{1}{R_2}$. Conductances in series are totaled by use of the formula $\frac{1}{G_T} = \frac{1}{G_1} + \frac{1}{G_2}$ which is the reciprocal of the series resistance total calculation $R_T = R_1 + R_2$. Conductance G is rated in mhos (℧). Resistance R is rated in ohms (Ω).

DIRECT CURRENT (DC) POWER FORMULAS

There are three basic direct current power formulas, which are as follows:

① $P = EI$ *WHERE:* E = voltage in volts

② $P = I^2R$ I = current in amperes

③ $P = \frac{E^2}{R}$ P = power in watts

R = resistance in ohms

Formula Variations of Electrical Power in Relation to Current and Voltage

BASE FORMULA: $P = EI$

LEGEND:
P = Power in Watts
E = Voltage in Volts
I = Current in Amperes

Direct Current Circuit Analysis

ANALYSIS:

① *If "P" is unknown*

   GIVEN: *THEN:* $P = EI$
   $E = 8$ volts $P = 8 \times 2$
   $I = 2$ amps $P = 16$ watts

② *If "I" is unknown*

   GIVEN: *THEN:* $I = \dfrac{P}{E}$

   $P = 16$ watts $I = \dfrac{16}{8}$

   $E = 8$ volts $I = 2$ amps

③ *If "E" is unknown*

   GIVEN: *THEN:* $E = \dfrac{P}{I}$

   $P = 16$ watts $E = \dfrac{16}{2}$

   $I = 2$ amps $E = 8$ volts

Formula Variations of Electrical Power in Relation to Current and Resistance

BASE FORMULA: $P = I^2R$

LEGEND:
P = Power in Watts
I = Current in Amperes
R = Resistance in Ohms

ANALYSIS:

① *If "P" is unknown*

   GIVEN: *THEN:* $P = I^2R$
   $I = 2$ amps $P = 2^2 \times 4$
   $R = 4$ ohms $P = 4 \times 4$
                                    $P = 16$ watts

② *If "I" is unknown*

   GIVEN: *THEN:* $I^2 = \dfrac{P}{R}$

   $P = 16$ watts $I^2 = \dfrac{16}{4}$

Basic Facts about Direct Current

$\quad\quad$ R = 4 ohms $\quad\quad\quad\quad$ $I^2 = 4$

$\quad\quad\quad\quad\quad\quad\quad\quad\quad\quad\quad\quad$ I = 2 amps

③ *If "R" is unknown*

$\quad\quad$ GIVEN: $\quad\quad\quad\quad$ THEN: $R = \dfrac{P}{I^2}$

$\quad\quad$ P = 16 watts $\quad\quad\quad\quad\quad\quad$ $R = \dfrac{16}{2^2}$

$\quad\quad$ I = 2 amps $\quad\quad\quad\quad\quad\quad$ $R = \dfrac{16}{4}$

$\quad\quad\quad\quad\quad\quad\quad\quad\quad\quad\quad\quad$ R = 4 ohms

Formula Variations of Power in Relation to Voltage and Resistance

BASE FORMULA: $P = \dfrac{E^2}{R}$

LEGEND:
P = Powers in Watts
E = Voltage in Volts
R = Resistance in Ohms

ANALYSIS:

① *If "P" is unknown*

$\quad\quad$ GIVEN: $\quad\quad\quad\quad$ THEN: $P = \dfrac{E^2}{R}$

$\quad\quad$ E = 8 volts $\quad\quad\quad\quad\quad\quad$ $P = \dfrac{8^2}{4}$

$\quad\quad$ R = 4 ohms $\quad\quad\quad\quad\quad\quad$ $P = \dfrac{64}{4}$

$\quad\quad\quad\quad\quad\quad\quad\quad\quad\quad\quad\quad$ P = 16 watts

② *If "E" is unknown*

$\quad\quad$ GIVEN: $\quad\quad\quad\quad$ THEN: $E^2 = RP$
$\quad\quad$ P = 16 watts $\quad\quad\quad\quad\quad\quad$ $E^2 = 4 \times 16$
$\quad\quad$ R = 4 ohms $\quad\quad\quad\quad\quad\quad$ $E^2 = 64$
$\quad\quad\quad\quad\quad\quad\quad\quad\quad\quad\quad\quad$ E = 8 volts

Direct Current Circuit Analysis

③ *If "R" is unknown*

GIVEN:　　　　　　　THEN: $R = \dfrac{E^2}{P}$

P = 16 watts　　　　　　　　$R = \dfrac{8^2}{16}$

E = 8 volts　　　　　　　　$R = \dfrac{64}{16}$

　　　　　　　　　　　　　R = 4 ohms

Section 2

Series DC Circuit

Series dc circuits are the most simple to analyze. All components of the series circuit are on one line from the power supply and return. The same current must flow through all components. Formulas are mostly simple additions. Basically, in a series dc circuit the voltage total is equal to the sum of the series voltage drops in the circuit. Current through a series circuit is the same throughout the circuit and through all of its components. Resistance total in a series circuit is the sum of the resistances in the circuit.

SERIES DC CIRCUIT FORMULAS (See Fig. 6.)

$E_T = E_1 + E_2$ *WHERE:* E_T = Total voltage in volts
$I_T = I_1 = I_2$ $\qquad\qquad\;\;$ E_1 = Voltage drop across first resistor
$R_T = R_1 + R_2$ $\qquad\;\;$ E_2 = Voltage drop across second resistor
$\qquad\qquad\qquad\qquad\;\;$ I_T = Total current in amperes
$\qquad\qquad\qquad\qquad\;\;$ I_1 = Current through first resistor
$\qquad\qquad\qquad\qquad\;\;$ I_2 = Current through second resistor
$\qquad\qquad\qquad\qquad\;\;$ R_T = Total resistance in ohms
$\qquad\qquad\qquad\qquad\;\;$ R_1 = Resistance of first resistor in ohms
$\qquad\qquad\qquad\qquad\;\;$ R_2 = Resistance of second resistor in ohms

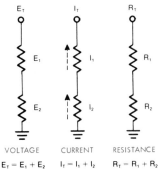

Fig. 6. Series DC Circuit Analysis

Formula Variations of Voltage Relationship in Series DC Resistive Circuits

BASE FORMULA: $E_T = E_1 + E_2 +$ etc.

LEGEND:

$E_T =$ Total voltage
$E_1 =$ Voltage drop, Resistor No. 1
$E_2 =$ Voltage drop, Resistor No. 2, etc.

ANALYSIS:

① *If "E_T" is unknown*
    GIVEN: *THEN:* $E_T = E_1 + E_2$
    $E_1 = 4$ volts $E_T = 4 + 2$
    $E_2 = 2$ volts $E_T = 6$ volts

② *If "E_1" is unknown*
    GIVEN: *THEN:* $E_1 = E_T - E_2$
    $E_T = 6$ volts $E_1 = 6 - 2$
    $E_2 = 2$ volts $E_1 = 4$ volts

③ *If "E_2" is unknown*
    GIVEN: *THEN:* $E_2 = E_T - E_1$
    $E_T = 6$ volts $E_2 = 6 - 4$
    $E_1 = 4$ volts $E_2 = 2$ volts

Formula Variations of Current Relationship in Series DC Resistive Circuits

BASE FORMULA: $I_T = I_1 = I_2 =$ etc.

LEGEND:

$I_T =$ Total current
$I_1 =$ Current through Resistor No. 1
$I_2 =$ Current through Resistor No. 2

ANALYSIS:

$I_T = 2$ amps
$I_1 = 2$ amps
$I_2 = 2$ amps

Formula Variations of Resistance Relationship in Series DC Resistive Circuits

FORMULA: $R_T = R_1 + R_2 +$ etc.
$R_T =$ Total resistance
$R_1 =$ Resistance of Resistor No. 1
$R_2 =$ Resistance of Resistor No. 2

ANALYSIS:

① *If "R_T" is unknown*
    GIVEN: THEN: $R_T = R_1 + R_2$
    $R_1 = 4$ ohms $R_T = 4 + 2$
    $R_2 = 2$ ohms $R_T = 6$ ohms

② *If R_1 is unknown*
    GIVEN: THEN: $R_1 = R_T - R_2$
    $R_T = 6$ ohms $R_1 = 6 - 2$
    $R_2 = 2$ ohms $R_1 = 4$ ohms

③ *If R_2 is unknown*
    GIVEN: THEN: $R_2 = R_T - R_1$
    $R_T = 6$ ohms $R_2 = 6 - 4$
    $R_1 = 4$ ohms $R_2 = 2$ ohms

VOLTAGE DIVISION (See Fig. 7.)

Consider the circuit of Fig. 7 as being a power supply with three separate voltage levels as pickoffs. The total resistance in this series circuit is the sum of 2K + 4K + 6K + 8K or 20K ohms.

Since there are 20 volts applied, by ratio there are 20V for 20K ohms resistance or 1 volt per 1 K ohms resistance. Therefore, from the point *A* looking to ground, there are 4K + 6K + 8K = 18K ohms resistance. At 1 volt per 1K ohm resistance, the voltage divided from point *A* to ground is 18V. From point *B* looking to ground there are 6K + 8K = 14K ohms resistance. At 1 volt per 1K ohm resistance, the voltage divided from point *B* looking to ground is 14 V. Finally from point *C* looking to ground there is 8K ohm resistance. At 1 volt per 1K ohm resistance, the voltage divided from point *C* looking to ground is 8 V.

Voltage division is most useful in any application where the exact

Direct Current Circuit Analysis

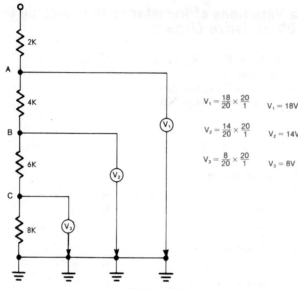

Fig. 7. Voltage Division

voltage needed is not available. A typical example is when you are using a single power supply to provide collector voltage and bias voltage for a transistor bias circuit.

In Fig. 8 this is illustrated. Note that the collector voltage re-

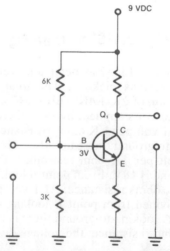

Fig. 8. Voltage Division Used for Transistor Biasing

Series DC Circuit

quired is 9 vdc, where the bias voltage required is 3 vdc. A simple voltage divider is used to supply two required voltages from the source.

In the figure a 6K resistor and a 3K resistor are used to voltage divide the power supply voltage to the 3 vdc necessary for the base of transistor Q1. You can see that the 6K + 3K equals 9K ohms resistance. Since there are 9 V applied, by ratio there are 9 V for 9K ohms resistance or 1 volt per 1K ohm resistance. Therefore from point *A* above the 3K resistor looking to ground, there is 3 V (at the rate of 1 volt per 1K ohm). The ratio of a voltage divider is not necessarily 1 to 1 as shown here. However, whether 1 to 1 or 100 to 1, the operation is the same.

"OPENS" IN A SERIES CIRCUIT (See Fig. 9.)

If we analyze the simple series circuit in Fig. 9(A) we see that there are 2 volts applied with 2 ohms of resistance, which causes 1 ampere current to flow.

If this circuit is open anywhere in the series path such as is shown in Fig. 9 (B), two situations will occur. First, the "open" will become an infinite resistance. Since this is the case, current flow will be so small that it will be non-measurable. The second situation that will occur is that all the voltage applied will be felt across the open. The

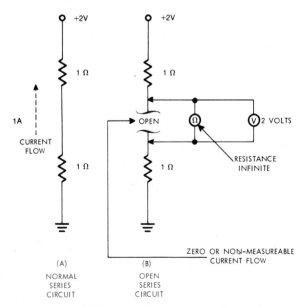

Fig. 9. Open in Series Circuit

Direct Current Circuit Analysis

reason for this is that the rest of the resistance in the circuit will become negligible compared to the infinite resistance of the "open." In theory there would be a voltage drop across the other resistors but too small for most test equipment to measure.

"SHORTS" IN A SERIES CIRCUIT (See Fig. 10.)

In a simple series circuit "shorts" can cause the problem of excessive current flow, which in turn will cause power to increase and circuit components to burn out.

In Fig. 10(A) the normal circuit is presented. The circuit has 6 volts and 2 amperes, therefore 12 watts (P = EI). In Fig. 10(B) the center resistance is shorted around. The short decreases the total resistance of the circuit from 3 ohms to 2 ohms, thereby driving the current up from 2 amperes to 3 amperes. Power in the circuit with this short increases from normal 12 watts to 18 watts.

In Fig. 10(C) the top resistance is shorted to ground. This effectively removes the other two resistances from the circuit. The short decreases the total resistance in the circuit from 3 ohms to 1 ohm and drives the current in the circuit up from 2 amperes to 6 amperes. Power in the circuit with this short increases from normal 12 watts to 36 watts.

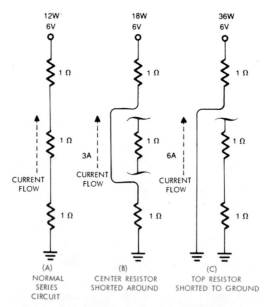

Fig. 10. Short in Series Circuits

Section 3

Parallel DC Circuit

Parallel dc circuits include as many parallel lines as desired. Equations require only simple addition and simple algebra. In a parallel circuit, voltage is the same throughout each parallel leg, no matter how many legs are added. Current through a parallel circuit is the sum of the currents in each leg. Resistance total calculations for a 2 resistor parallel circuit is the product of the resistors divided by the sum of the resistors. Calculations for more than 2 resistors are shown in Fig. 12 and described under a separate subheading.

PARALLEL DC CIRCUIT FORMULAS (See Fig. 11.)

$$E_T = E_1 = E_2$$
$$I_T = I_1 + I_2$$
$$R_T = \frac{R_1 R_2}{R_1 + R_2}$$

R_T *for More Than 2 Resistors:*

$$R_T = \frac{1}{\frac{1}{R_1} + \frac{1}{R_2} + \frac{1}{R_N}} \quad \text{or} \quad \frac{1}{R_T} = \frac{1}{R_1} + \frac{1}{R_2} + \frac{1}{R_N}$$

Note: R_N *represents any number of resistors.*

Direct Current Circuit Analysis

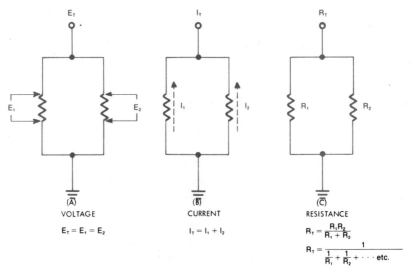

Fig. 11. Parallel DC Circuit Analysis

Formula Variations of Voltage Relationship in Parallel DC Resistive Circuits

BASE FORMULA: $E_T = E_1 = E_2 =$ etc.

LEGEND:
$E_T =$ Total voltage
$E_1 =$ Voltage drop, Resistor No. 1
$E_2 =$ Voltage drop, Resistor No. 2, etc.
 Any number of resistances may be added

ANALYSIS:
$E_T = 2$ volts
$E_1 = 2$ volts
$E_2 = 2$ volts

Note: *Voltage remains the same on each leg of a parallel circuit no matter how many resistances are added.*

Formula Variations of Current Relationship in Parallel DC Resistive Circuits

BASE FORMULA: $I_T = I_1 + I_2 +$ etc.

22

LEGEND:
I_T = Total current
I_1 = Current through Resistor No. 1
I_2 = Current through Resistor No. 2
  Any number of currents may be added

ANALYSIS:

① *If "I_T" is unknown*

  GIVEN: THEN: $I_T = I_1 + I_2$
  $I_1 = 2$ amps $I_T = 2 + 4$
  $I_2 = 4$ amps $I_T = 6$ amps

② *If "I_1" is unknown*

  GIVEN: THEN: $I_1 = I_T - I_2$
  $I_T = 6$ amps $I_1 = 6 - 4$
  $I_2 = 4$ amps $I_1 = 2$ amps

③ *If "I_2" is unknown*

  GIVEN: THEN: $I_2 = I_T - I_1$
  $I_T = 6$ amps $I_2 = 6 - 2$
  $I_1 = 2$ amps $I_2 = 4$ amps

Formula Variations of Resistance Relationship in Parallel DC Resistive Circuits (2 Resistors)

BASE FORMULA: $R_T = \dfrac{R_1 R_2}{R_1 + R_2}$

LEGEND:
R_T = Total resistance
R_1 = Resistance of Resistor No. 1
R_2 = Resistance of Resistor No. 2

ANALYSIS:

① *If "R_T" is unknown*

  GIVEN: THEN: $R_T = \dfrac{R_1 R_2}{R_1 + R_2}$

  $R_1 = 3$ ohms $R_T = \dfrac{3 \times 6}{3 + 6}$

  $R_2 = 6$ ohms $R_T = \dfrac{18}{9}$

                               $R_T = 2$ ohms

Direct Current Circuit Analysis

② *If "R_1" is unknown*

GIVEN: THEN: $R_1 = \dfrac{R_2 R_T}{R_2 - R_T}$

$R_T = 2$ ohms

$R_1 = \dfrac{6 \times 2}{6 - 2}$

$R_2 = 6$ ohms

$R_1 = \dfrac{12}{4}$

$R_1 = 3$ ohms

③ *If "R_2" is unknown*

GIVEN: THEN: $R_2 = \dfrac{R_1 R_T}{R_1 - R_T}$

$R_T = 2$ ohms

$R_2 = \dfrac{3 \times 2}{3 - 2}$

$R_1 = 3$ ohms

$R_2 = \dfrac{6}{1}$

$R_2 = 6$ ohms

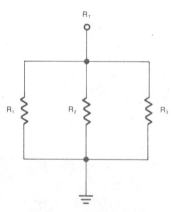

RESISTANCE [3 OR MORE RESISTORS]

$\dfrac{1}{R_T} = \dfrac{1}{R_1} + \dfrac{1}{R_2} + \dfrac{1}{R_3}$ or $R_T = \dfrac{1}{\dfrac{1}{R_1} + \dfrac{1}{R_2} + \dfrac{1}{R_3}}$

Fig. 12. Resistance Totals in a Parallel Circuit with 3 or more Resistors

Parallel DC Circuit

Formula Variations of Resistance Relationship in Parallel DC Resistive Circuits (3 or More Resistors) (See Fig. 12.)

BASE FORMULA: $\dfrac{1}{R_T} = \dfrac{1}{R_1} + \dfrac{1}{R_2} +$ etc.

LEGEND:
R_T = Total resistance
R_1 = Resistance of Resistor No. 1
R_2 = Resistance of Resistor No. 2
R_3 = Resistance of Resistor No. 3
  Any number of resistors may be added

ANALYSIS:

① If "R_T" is unknown

GIVEN: THEN: $\dfrac{1}{R_T} = \dfrac{1}{R_1} + \dfrac{1}{R_2} + \dfrac{1}{R_3}$

$R_1 =$ 8 ohms $\dfrac{1}{R_T} = \dfrac{1}{8} + \dfrac{1}{12} + \dfrac{1}{24}$

$R_2 =$ 12 ohms $\dfrac{1}{R_T} = \dfrac{3}{24} + \dfrac{2}{24} + \dfrac{1}{24}$

$R_3 =$ 24 ohms $\dfrac{1}{R_T} = \dfrac{6}{24}$

                            $R_T = \dfrac{24}{6} = 4$ ohms

② If "R_1" is unknown

GIVEN: THEN: $\dfrac{1}{R_T} = \dfrac{1}{R_1} + \dfrac{1}{R_2} + \dfrac{1}{R_3}$

$R_T =$ 4 ohms $\dfrac{1}{4} = \dfrac{1}{R_1} + \dfrac{1}{12} + \dfrac{1}{24}$

$R_2 =$ 12 ohms $-\dfrac{1}{R_1} = \dfrac{2}{24} + \dfrac{1}{24} - \dfrac{6}{24}$

$R_3 =$ 24 ohms $-\dfrac{1}{R_1} = -\dfrac{3}{24}$

                            $R_1 = \dfrac{24}{3} = 8$ ohms

If "R_2" or "R_3" are unknown, procedures are the same as in R_1 calculation procedures.

"OPENS" IN A PARALLEL CIRCUIT (See Fig. 13.)

In analyzing the simple parallel circuit in Fig. 13(A) we see that there are 6 volts applied, with 0.66 ohms total resistance, which causes 9 amperes of current to flow.

If this circuit is opened as shown in Fig. 13(B), several situations will occur. First, the open will become an infinite resistance. Since this is the case, current flow through this particular leg will be essentially zero. The second situation that will occur is that all the voltage drop on that open leg will be across the "open" in the circuit. The reason for this is, that the other resistance in the leg will become negligible compared to the infinite resistance of the "open." In theory, there would be a voltage drop across the resistor, but too small for test equipment to measure. The third situation that will occur is that the total resistance of the entire circuit will increase from 0.66 ohms to 1 ohm. This effect will lower the current from 9 amperes to 6 amperes.

This same situation occurs in a house circuit when a light bulb burns out. The burnt filament in the bulb becomes the open.

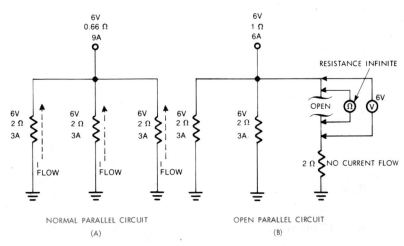

Fig. 13. Open in a Parallel Circuit

"SHORTS" IN A PARALLEL CIRCUIT (See Fig. 14.)

In a simple parallel circuit, shorts can cause the problem of excessive current flow through the shorted leg, which normally causes

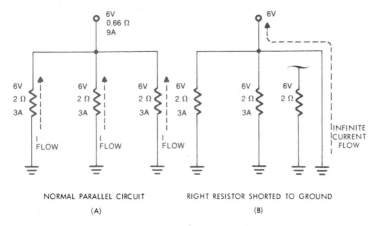

Fig. 14. Short in a Parallel Circuit

the component on that leg to burn out. If the short circuit contains only wire, the wire will normally get hot and burn open.

In Fig. 14(A) the normal circuit is presented. In Fig. 14(B) the right resistor is shorted out. The connecting wire is shorted to ground.

Current through the shorted wire increases in direct relation to the resistance of the wire. If the wire is made of copper, for instance, the resistance per 1000 feet of number 10 wire is approximately 1 ohm. A one-foot piece of copper wire would have 0.001 ohm resistance.

If 6 volts were applied to 0.001 ohm wire resistance, as in our example in Fig. 14(B), the current flow through the wire could literally be 6000 amperes if the power source had an infinite capacity. This would immediately melt the wire in two. There would be no current flow in the other two resistance legs because the draw of current for the short is too great. Current takes the easiest path for flow. Consequently, the other two legs would not be affected by the short. After the shorted leg burns in two, current will again flow in the other two legs at the original level.

Section 4

The Wheatstone Bridge

The Wheatstone bridge (Fig. 15) is used to determine the resistance of any unknown resistor. In Fig. 15, R_1 and R_2 are a voltage divider from the power source. R_{UNK} (the unknown resistance) and R_{VAR} (variable calibrated resistance) are the second voltage divider from the power supply. These resistances are all very accurate components.

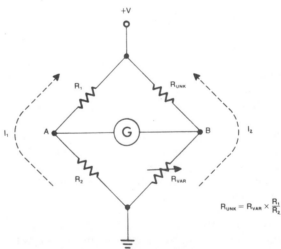

Fig. 15. The Wheatstone Bridge

The Wheatstone Bridge

A precision galvanometer is placed between points A and B. When power is applied to the circuit, the galvanometer will indicate current flow in either the point A or point B direction. Resistor R_{VAR} is then varied until the galvanometer indicates no current flow. This shows that the bridge is balanced. When balanced, the bridge sets up a mathematical voltage ratio between the two voltage dividers. This is shown as follows:

$$\frac{I_1 R_1}{I_1 R_2} = \frac{I_2 R_{UNK}}{I_2 R_{VAR}}$$

The current component can be removed from the equation because current is equal in each leg. The ratio then reads:

$$\frac{R_1}{R_2} = \frac{R_{UNK}}{R_{VAR}} \quad \text{or} \quad R_{UNK} = R_{VAR} \times \frac{R_1}{R_2}$$

This same bridge is used extensively in automatic control systems and also electrical measuring instruments.

Section

Series Parallel DC Circuit

Series parallel circuits are complex circuits composed of both series and parallel legs. Electronic circuits are in general made up of these complex circuits. Because electronic circuits normally utilize a single power supply, voltage must be series-divided to provide dc levels of various magnitudes for circuit operation. In the same circuits, each stage of electronics is a parallel leg. It is therefore imperative that the analysis of series parallel circuits be high on your list of electronic learning objectives.

VOLTAGE IN A SERIES PARALLEL DC CIRCUIT
(See Fig. 16.)

In a series parallel dc circuit, total voltage is calculated by first drawing an equivalent circuit as in Fig. 16(B). Note that the parallel circuit for E_2 and E_3 can be represented by E_4 since E_2 and E_3 are in parallel, therefore are equal. The total voltage then, as shown in Fig. 16(B), is the sum of the voltage drops in the series circuit: $E_T = E_1 + E_4$.

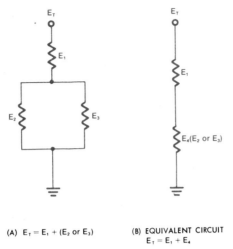

(A) $E_T = E_1 + (E_2 \text{ or } E_3)$ (B) EQUIVALENT CIRCUIT
$E_T = E_1 + E_4$

Fig. 16. Total Voltage Calculations in a Series Parallel DC Circuit

Example of Voltage Calculation for Series Parallel DC Circuit (See Fig. 16.)

GIVEN:
$E_1 = 2$ volts
$E_2 = 4$ volts
$E_3 = 4$ volts
$E_4 = E_2 = E_3$
$E_4 = 4$ volts
Then in the equivalent circuit, Fig. 16(B),
$E_T = E_1 + E_4$
$E_T = 2 + 4$
$E_T = 6$ volts

RESISTANCE IN A SERIES PARALLEL DC CIRCUIT (See Fig. 17.)

In a series parallel dc circuit, solution requires that the parallel paths be solved initially, then an equivalent series circuit be drawn. In Fig. 17 the resistors R_2 and R_3 are converted to resistor R_5 using the total resistance formula $R_T = \dfrac{R_2 R_3}{R_2 + R_3}$.

After conversion, the equivalent series circuit is analyzed in the same manner as ordinary series dc circuits.

Example of Resistance Calculation for Series Parallel Circuit (See Fig. 17.)

GIVEN:

$R_1 = 5$ ohms

$R_2 = 10$ ohms

$R_3 = 40$ ohms

$R_4 = 80$ ohms

THEN: $R_5 = \dfrac{R_2 R_3}{R_2 + R_3}$

$R_5 = \dfrac{10 \times 40}{10 + 40}$

$R_5 = \dfrac{400}{50}$

$R_5 = 8$ ohms

Then in the equivalent circuit, Fig. 17(B),
$R_T = R_1 + R_5 + R_4$
$R_T = 5 + 8 + 80$
$R_T = 93$ ohms

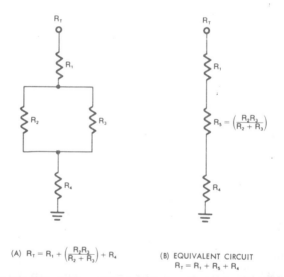

(A) $R_T = R_1 + \left(\dfrac{R_2 R_3}{R_2 + R_3}\right) + R_4$

(B) EQUIVALENT CIRCUIT
$R_T = R_1 + R_5 + R_4$

Fig. 17. Total Resistance Calculations in a Series Parallel DC Circuit

CURRENT IN A SERIES PARALLEL DC CIRCUIT
(See Fig. 18.)

In a series parallel dc circuit, total current can be calculated in two ways. First, by calculating the total resistance, then using

Series Parallel DC Circuit

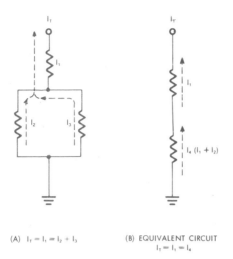

(A) $I_T = I_1 = I_2 + I_3$ (B) EQUIVALENT CIRCUIT
$I_T = I_1 = I_4$

Fig. 18. Total Current Calculations in a Series Parallel DC Circuit

Ohm's law to calculate current. Secondly, current can be calculated as shown in Fig. 18. Note that the two currents I_2 and I_3 have to be added together as in the equivalent circuit of Fig. 18(B). Then the total current is calculated as in any series circuit.

Example of Current Calculation for Series Parallel DC Circuit (See Fig. 18(A).)

GIVEN:
$I_2 = 10$ amps
$I_3 = 15$ amps

THEN:
$I_4 = I_2 + I_3$
$I_4 = 10 + 15$
$I_4 = 25$ amps

Then in the equivalent circuit, Fig. 18(B),
$I_T = I_1 = I_4$

$I_T = 25$ amps, or $I_T = \dfrac{E_T}{R_T}$

VOLTAGE DIVISION IN A COMPLEX CIRCUIT
(See Fig. 19.)

In Fig. 19 there are several parallel paths. Analyses of these paths are from any point looking at the path or paths to ground.

Direct Current Circuit Analysis

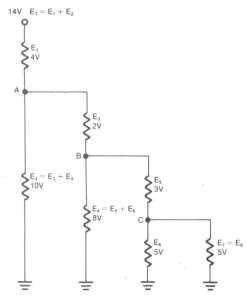

Fig. 19. Voltage Division in a Complex Circuit

① From point C, E_6 is in parallel with E_7.

② From point B, E_4 is in parallel with $E_5 + E_6$ and also in parallel with $E_5 + E_7$.

③ From point A, E_2 is in parallel with $E_3 + E_4$, also in parallel with $E_3 + E_5 + E_6$, and finally in parallel with $E_3 + E_5 + E_7$.

Each parallel branch is equal in voltage value.

CALCULATING TOTAL RESISTANCE IN A COMPLEX CIRCUIT (See Fig. 20.)

The method most utilized for calculating total resistance in a complex circuit is called the *equivalent circuit method*. The philosophy is to reduce the complex circuit to an equivalent simple series circuit. Calculation, then, is simple addition.

In Fig. 20(A) the basic circuit is illustrated. The procedure for reducing the circuit is as follows:

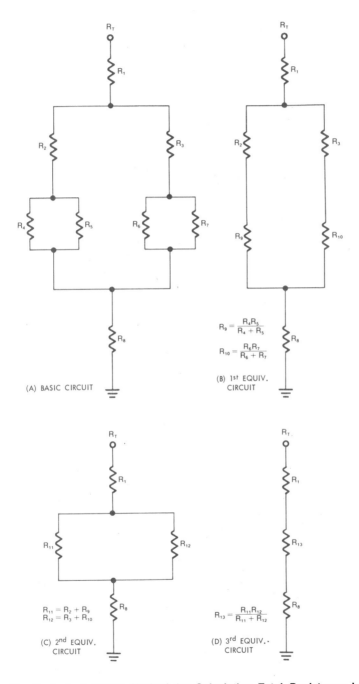

Fig. 20. Equivalent Circuit Method for Calculating Total Resistance in a Complex Circuit

Direct Current Circuit Analysis

① Reduce R_4 and R_5 to a single series resistance R_9 as in Fig. 20(B).

$$R_9 = \frac{R_4 \times R_5}{R_4 + R_5}$$

② Reduce R_6 and R_7 to a single series resistance R_{10} as in Fig. 20(B).

$$R_{10} = \frac{R_6 \times R_7}{R_6 + R_7}$$

③ Reduce R_2 and R_9 to a single series resistance R_{11} as in Fig. 20(C).

$$R_{11} = R_2 + R_9$$

④ Reduce R_3 and R_{10} to a single series resistance R_{12} as in Fig. 20(C).

$$R_{12} = R_3 + R_{10}$$

⑤ Reduce R_{11} and R_{12} to a single series resistance R_{13} as in Fig. 20(D).

$$R_{13} = \frac{R_{11} \times R_{12}}{R_{11} + R_{12}}$$

⑥ Calculate $R_T = R_1 + R_{13} + R_8$

Section 6

Kirchoff's Laws for Current and Voltage

KIRCHOFF'S CURRENT LAW (See Fig. 21.)

Kirchoff's current law states that the algebraic sum of currents entering a junction is equal to the current leaving the junction. Another way of saying this is that there is exactly as much current leaving a junction as there is entering the junction.

Looking at Fig. 21, you will note that the current flowing toward the junction is labeled I_1 and I_2, while current flowing away from the junction is labeled I_3.

Example of Current Calculation Using Kirchoff's Current Law (See Fig. 21.)

GIVEN:
$I_1 = 5$ amps
$I_2 = 3$ amps
$I_3 = 8$ amps

THEN:
$I_1 + I_2 = I_3$
$5 + 3 = 8$ amps

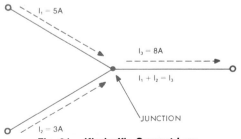

Fig. 21. Kirchoff's Current Law

Direct Current Circuit Analysis

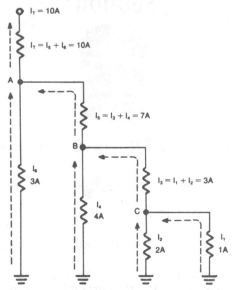

Fig. 22. Kirchoff's Current Law in a Complex Circuit

Kirchoff's Current Law in a Complex Circuit (See Fig. 22.)

In Fig. 22, current through I_1 plus I_2 equals the current through I_3. Similarly, the current through I_3 plus I_4 equals the current through I_5. Finally, the current through I_5 plus I_6 equals the current thru I_T.

① From point C, $I_3 = I_1 + I_2$

② From Point B, $I_5 = I_3 + I_4$

③ From point A, $I_T = I_5 + I_6$

KIRCHOFF'S VOLTAGE LAW (See Fig. 23.)

Kirchoff's voltage law states that the algebraic sum of voltage inputs and voltage drops in a closed loop are equal to zero. In Fig. 23, if you start at point A and go counterclockwise around the closed loop back to point A, the voltage gains and drops are equal to zero.

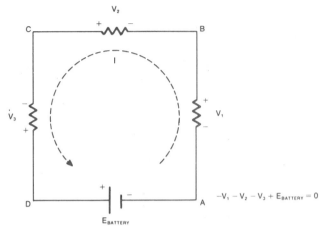

Fig. 23. Kirchoff's Voltage Law

Example of Voltage Calculation Using Kirchoff's Voltage Law (See Fig. 23.)

GIVEN:
 $V_1 = 4$ volts
 $V_2 = 2$ volts
 $V_3 = 3$ volts
 $E_{BAT} = 9$ volts

THEN:
 CCW* from point A
 $-V_1 - V_2 - V_3 + E_{BAT} = 0$
 $-4 - 2 - 3 + 9 = 0$
 $0 = 0$
 *CCW = Counterclockwise

Section 7

Mesh Circuits

Mesh circuits (Fig. 24) are a series of circuit loops which have one or more common elements. The mesh loop is the smallest loop in a network. Analysis begins at any junction. This junction can be any point in the loop.

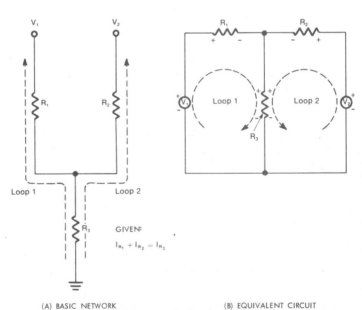

(A) BASIC NETWORK (B) EQUIVALENT CIRCUIT

Fig. 24. Mesh Circuit Analysis

MESH CIRCUIT ANALYSIS

Negative and positive polarities for each resistance should be assigned prior to analysis. These polarities are assigned in relation to the negative and positive terminals of the battery.

From the junction a continuous closed path is followed through the loop without passing through another element more than once, before returning to the beginning junction. This current path is called a current loop. A network having two or more such loops with common elements is a candidate for mesh analysis.

In Fig. 24 the resistor R_3 is the common element to the two current loops one (1) and two (2).

The mesh network currents are calculated by solving a set of simultaneous equations. As a rule of thumb, an equation must be written for each current loop, while strictly observing polarities within the mesh. The equations are then solved simultaneously.

Mesh Circuit Calculation (Procedure)

(1) Write an equation for loop 1 using Kirchoff's voltage law.

$$IR_1 + IR_3 - V_1 = 0$$

(2) Write an equation for loop 2 using Kirchoff's voltage law.

$$IR_2 + IR_3 - V_2 = 0$$

(3) Solve the equations (1) and (2) using the simultaneous equation method.

Note: $I_{R_1} + I_{R_2} = I_{R_3}$

Mesh Circuit Calculation (Example)

GIVEN:
$R_1 = 3$ ohms
$R_2 = 6$ ohms
$R_3 = 4$ ohms
$V_1 = 29$ volts
$V_2 = 32$ volts
Note: $I_3 = I_1 + I_2$

(1) $3(I_1) + 4[I_1 + I_2] - 29 = 0$

(2) $6(I_2) + 4[I_1 + I_2] - 32 = 0$

REMOVE BRACKETS:

(1) $3(I_1) + 4(I_1) + 4(I_2) - 29 = 0$

(2) $6(I_2) + 4(I_1) + 4(I_2) - 32 = 0$

Direct Current Circuit Analysis

COMBINE:

① $7(I_1) + 4(I_2) = 29$
② $4(I_1) + 10(I_2) = 32$

Note: *In order to isolate one unknown, the coefficients of one of the variable quantities must have equal absolute value in both equations. If necessary, one or both equations may be multiplied by any number that will satisfy this requirement.*

MULTIPLY ① BY 4 AND ② BY −7

① $28(I_1) + 16(I_2) = 116$
② $-28(I_1) - 70(I_2) = -224$

ALGEBRAICALLY ADD
$-54(I_2) = -108$
Solve for I_2
$I_2 = 2$ amps

SOLVE FOR I_1 USING EQUATION ①
$7(I_1) + 4(I_2) = 29$
$7(I_1) + 4(2) = 29$
$7(I_1) + 8 = 29$
$7(I_1) = 21$
$I_1 = 3$ amps

Section

Superposition Principle

Superposition is an extremely valuable tool when called on to solve resistance circuits that are excited with multiple voltage sources. The superposition theorem states that the response of a circuit containing two or more voltage sources is the algebraic sum of the responses of the individual voltage sources.

METHOD OF CALCULATION BY SUPERPOSITION PRINCIPLE

The method of calculation is to analyze the response of a single source with the other sources as zero impedance. In Fig. 25(A) a basic circuit is shown. In Fig. 25(B) the voltage source V_2 is shorted to ground. The circuit is analyzed with the single voltage source V_1, and the equivalent resistor R_4. In Fig. 25(C) the voltage source V_1 is shorted to ground and the circuit is analyzed with the single voltage source V_2 and the equivalent resistor R_5.

Current calculations are performed for the common resistor R_3 in each case and then these currents are summed to find the total current, I_{R_3} applying Kirchoff's current law.

Calculations for Figs. 25(A) through 25(C) assume each voltage source to be ideal (with zero source impedance). If source impedance is significant, it should be substituted for the voltage source that is shorted in Figs. 25(B) and 25(C) and included in the calculations.

Direct Current Circuit Analysis

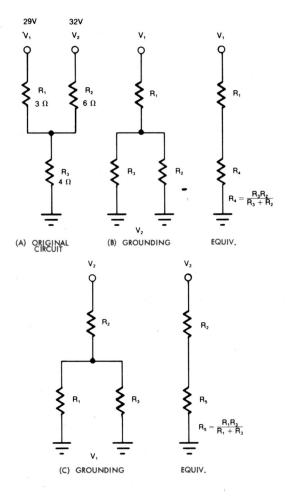

Fig. 25. Superposition Principle

Superposition Principle (Procedure)

OBJECTIVE: Find current I_{R_3} and voltage E_{R_3} (common resistor)

① Short out V_2 as shown in Figure 25(B)

② Calculate $R_{T1} = R_1 + \dfrac{R_2 R_3}{R_2 + R_3}$ since $R_4 = \dfrac{R_2 R_3}{R_2 + R_3}$

③ Short out V_1 as shown in Figure 25(C)

④ Calculate $R_{T_2} = R_2 + \dfrac{R_1 R_3}{R_1 + R_3}$ since $R_5 = \dfrac{R_1 R_3}{R_1 + R_3}$

⑤ Calculate $I_{T_1} = \dfrac{V_1}{R_{T_1}}$

⑥ Calculate $I_{T_2} = \dfrac{V_2}{R_{T_2}}$

⑦ Calculate $E_{R_4} = I_{T_1} R_4$ (E_{R_4} is that part of V_1 that is across R_3.)

⑧ Calculate $E_{R_5} = I_{T_2} R_5$ (E_{R_5} is that part of V_2 that is across R_3.)

⑨ Calculate $I_{R_3} = \dfrac{E_{R_4}}{R_3} + \dfrac{E_{R_5}}{R_3} = \dfrac{E_{R_4} + E_{R_5}}{R_3}$

⑩ Calculate $E_{R_3} = I_{R_3} \times R_3$

Superposition Principle (Example)

GIVEN:
$R_1 = 3$ ohms $R_3 = 4$ ohms $V_1 = 29$ volts
$R_2 = 6$ ohms $V_2 = 32$ volts

FIND: I_{R_3} and E_{R_3}

① Short V_2

② $R_{T_1} = 3 + \dfrac{6 \times 4}{6 + 4} = 3 + \dfrac{24}{10} = 3 + 2.4 = 5.4$ ohms

③ Short V_1

④ $R_{T_2} = 6 + \dfrac{3 \times 4}{3 + 4} = 6 + \dfrac{12}{7} = 6 + 1.71 = 7.71$ ohms

⑤ $I_{T_1} = \dfrac{29}{5.4} = 5.37$ amps

⑥ $I_{T_2} = \dfrac{32}{7.71} = 4.15$ amps

⑦ $E_{R_4} = 5.37 \times 2.4 = 12.89$ volts

⑧ $E_{R_5} = 4.15 \times 1.71 = 7.10$ volts

⑨ $I_{R_3} = \dfrac{12.89 + 7.10}{4} = \dfrac{19.99}{4} = 5$ amps

⑩ $E_{R_3} = 5 \times 4 = 20$ volts

Section

Thévenin's Theorem

In the event power, current, or voltage value of a specific circuit component is desired, it may be convenient to use one of several theorems. One of these that is especially useful is Thévenin's Theorem. For purposes of calculation, Thévenin's Theorem allows a normally complex circuit with linear components to be simplified to a series circuit with an ideal voltage source and a series resistance. Linear components are components that allow current to increase proportionately to an increase in voltage.

From the (simplified) equivalent circuit it can easily be determined what load resistance would produce maximum power transfer and what the maximum realizable voltage across the load is. This information is very helpful when matching amplifiers, etc.

METHOD OF CALCULATION USING THÉVENIN'S THEOREM

Essentially, the theorem procedure is to disconnect the unknown component (R_L), reduce the rest of the circuit (R_1 and R_2) to a series equivalent resistor (R_{THEV}) by shorting voltage sources and opening current sources, then insert the unknown component into the simplified reduction for calculations.

Thévenin's Theorem (Procedure) (See Fig. 26.)

OBJECTIVE: Find current I_{THEV} and voltage E_{RL} (load resistance)

(1) Disconnect the load R_L from the circuit (Fig. 26(B).)

(2) With the load terminals open, calculate E_{THEV} by any convenient method. In this particular case, calculate by ratio method (Fig. 26(B).)

$$E_{THEV} = \frac{R_2}{R_1 + R_2} \times E_T$$

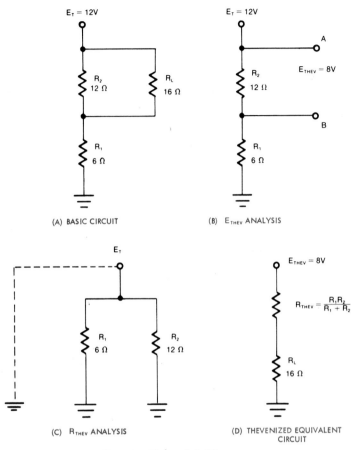

Fig. 26. Thévenin's Theorem

③ Short out power source and calculate R_{THEV}, as seen at open load terminals by parallel resistance calculation (Fig. 26(C).)

$$R_{THEV} = \frac{R_1 R_2}{R_1 + R_2}$$

④ Thévenize the circuit as in Fig. 26(D), using E_{THEV} and R_{THEV}.

⑤ Replace load R_L in Thévenized series circuit.

⑥ Calculate $I_{THEV} = \dfrac{E_{THEV}}{R_{THEV} + R_L}$

⑦ Calculate load voltage E_{RL} using Ohm's Law.

$$E_{RL} = I_{THEV} R_L$$

Direct Current Circuit Analysis

Thévenin's Theorem (Example)

GIVEN:
$R_1 = 6$ ohms $E_T = 12$ V
$R_2 = 12$ ohms
$R_L = 16$ ohms

FIND: I_{THEV} and E_{RL}

① Disconnect R_L

② $E_{THEV} = \dfrac{R_2}{R_1 + R_2} \times E_T$

$E_{THEV} = \dfrac{12}{6 + 12} \times 12$

$E_{THEV} = 8$ volts

③ $R_{THEV} = \dfrac{R_1 R_2}{R_1 + R_2}$

$R_{THEV} = \dfrac{6 \times 12}{6 + 12}$

$R_{THEV} = 4$ ohms

④ Draw equivalent circuit (Fig. 26(D).)

⑤ Replace R_L in equivalent circuit (Fig. 26(D).)

⑥ I_{THEV} (therefore I_{RL}) $= \dfrac{E_{THEV}}{R_{THEV} + R_L}$

$I_{THEV} = \dfrac{8}{4 + 16}$

$I_{THEV} = 0.4$ amp

⑦ $E_{RL} = I_{THEV} R_L$
$E_{RL} = 0.4 \times 16$
$E_{RL} = 6.4$ volts

Note: *To prove I_{THEV} and E_{RL} these values may be inserted into the original circuit.*

Section 10

Norton's Theorem

Norton's theorem is similar in function to Thévenin's theorem. Its use allows reduction of complex driving networks with linear components to an ideal current source in parallel with a resistor and simplifies voltage, current, and power calculations in the load area. Norton's theorem is most useful in applications that have a constant current source, but in practice is often used with Thévenin's theorem to reduce networks. The two theorems are complementary.

As an example, Norton's theorem may be applied to the basic circuit shown in Fig. 27(A) or 27(B). The objective is to find the current flow through the load resistor R_L.

METHOD OF CALCULATION USING NORTON'S THEOREM

The procedure for calculation is to first short out the load resistor as shown in Fig. 27(C) and find the current I_{NORT}. Next, as shown in Fig. 27(D), the total resistance is calculated. This resistance total is found by Thevenizing the circuit for R_{TOTAL}, which is equal to R_{THEV} and also R_{NORT}. Finally, current through the load resistor (I_{RL}) is found by using I_{NORT} and R_{NORT} previously calculated.

Norton's Theorem Procedures

OBJECTIVE: Find current flow through R_L (load resistance)

FIND: I_{R_L} (See Fig. 27(A).)

① Calculate $I_{R_3} = \dfrac{V}{R_3 + \dfrac{R_2 R_1}{R_2 + R_1}}$ (See Fig. 27(C).)

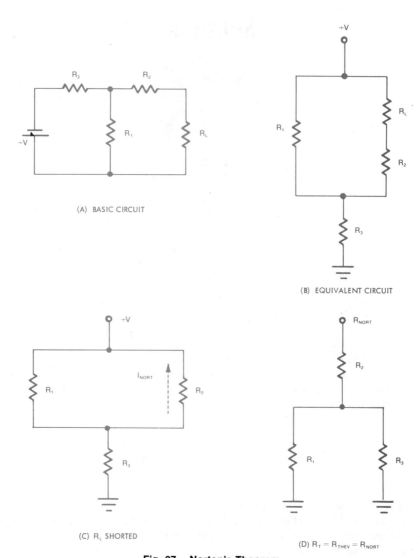

Fig. 27. Norton's Theorem

② Calculate $I_{NORT} = \dfrac{R_1}{R_2 + R_1} \times I_{R_3}$ (See Fig. 27(C).)

③ Calculate $R_{NORT} = R_{THEV} = R_2 + \dfrac{R_1 R_3}{R_1 + R_3}$ (See Fig. 27(D).)

④ Calculate $I_{R_L} = \dfrac{R_{NORT}}{R_{NORT} + R_L} \times I_{NORT}$ (See Fig. 27(A).)

Norton's Theorem (Example)

GIVEN:
$R_1 = 14$ ohms
$R_2 = 10$ ohms
$R_3 = 7$ ohms

$R_L = 6$ ohms
$V = 6$ volts

FIND: I_{R_L}

① $I_{R_3} = \dfrac{6}{7 + \dfrac{10 \times 14}{10 + 14}} = \dfrac{6}{7 + 5.83} = \dfrac{6}{12.83} = 0.47$ amp

② $I_{NORT} = \dfrac{14}{10 + 14} \times 0.47 = 0.27$ amp

③ $R_{NORT} = 10 + \dfrac{14 \times 7}{14 + 7} = 10 + 4.67 = 14.67$ ohms

④ $I_{R_L} = \dfrac{14.67}{14.67 + 6} \times 0.27 = 0.71 \times 0.27 = 0.19$ amp

Section 11

Equivalent Circuits

It is often useful to be able to redraw a complex circuit into a simplified circuit that is equivalent. Suppose our need is to know the current, voltages, and/or power in one specific branch of a circuit. To find these, we require only two bits of information about the remainder of the circuit: (a) How much current can the branch draw from the rest of the circuit if the branch resistance is reduced to zero (the short circuit current)? (b) How much voltage will the circuit provide across the branch if the branch resistance is raised to infinity?

METHOD OF CALCULATING AN EQUIVALENT CIRCUIT

All circuits that provide the same short circuit current and the same open circuit voltages are equivalent. Two simplified circuits provide these needs: (a) An ideal voltage source in series with one resistor (Thévenin's Equivalent); and (b) An ideal current source in parallel with one resistor (Norton's Equivalent). The ideal voltage source is equal to the open circuit voltage. The ideal current source is equal to the short circuit current. The series resistor and the parallel resistor are equal, and have a value of V open ÷ by I short.

Conversion Between Thévenin and Norton

Source transformations aid in solving some circuits for their Thévenin or Norton equivalents. In the example shown in Fig. 28(A), Thévenin's Theorem provides no advantage to be gained by source transformations as far as load (R_L) current is concerned. However, having obtained the Thévenin equivalent, (see Fig. 28(A),

Equivalent Circuits

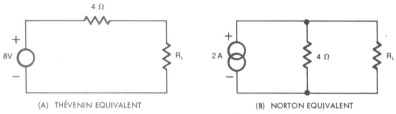

Fig. 28. Conversion between Thévenin and Norton

the Norton equivalent easily follows by shorting the load resistor R_L to determine I_{NORT}. Then place the Thévenin series resistance in parallel to the current source I_{NORT}, since $R_{NORT} = R_{THEV}$. (See Fig. 28(B).)

Conversion Between Norton and Thévenin by Source Transformations

In some instances, a circuit can be reduced to a Thévenin or Norton equivalent in less time by repeated source transformations. The basic circuit of Fig. 29(A) is transformed in several stages.

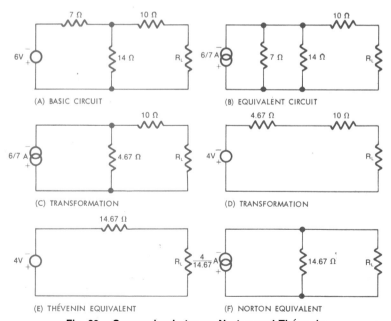

Fig. 29. Conversion between Norton and Thévenin

First, the voltage source 6 V and the 7 ohm resistor of Fig. 29(A) is converted to a current source $6/7$ A, as shown in Fig. 29(B). The 7 ohm and the 14 ohm resistors can then be parallel calculated to obtain 4.67 ohms, as shown in Fig. 29(C). Next, the current source $6/7$ A and the 4.67 ohm resistor are transformed back to an equivalent voltage source 4 V and series 4.67 ohm resistor as illustrated in Fig. 29(D). Combining of the 4.67 ohm and 10 ohm resistors produces the Thévenin equivalent circuit (14.67 ohms) as shown in Fig. 29(E). Fig. 29(F) shows the Norton equivalent circuit.

It is immediately obvious from these two equivalent circuits, what voltage and current maximums can be supplied to the load. Since maximum power is obtained at $1/2$ maximum voltage and current, power is calculated by $1/4$ of the product of Thévenin's voltage and Norton's current.

Section 12

Millman's Theorem

Millman's theorem is a variation of Norton's theorem. It is used to find the voltage value between two points in a network. In Fig. 30(A) the basic circuit is segregated into sections. Each section is then Nortonized, to find the Norton equivalent current of each section, as in Fig. 30(B).

The reader will note that since there is no power source in section II, there is no Norton equivalent current for the leg. In Fig. 30(C) the Norton equivalent currents are added and resistance is totaled, using standard parallel circuit analysis.

METHOD OF CALCULATION USING MILLMAN'S THEOREM

In Fig. 30(D) Millman's theorem is shown. The voltage value between point X and Y in the figure is found by multiplying the total Norton equivalent current by the total resistance R_T which in dc circuits is the reciprocal of conductance $\left(\frac{1}{G}\right)$. We recall that the reciprocal of resistance (R) in dc circuits is conductance (G), or $G = \frac{1}{R_T}$. In ac circuits, resistance is known as *impedance* (Z), and its reciprocal is *admittance* (Y). In ac circuits, as in dc circuits, admittance is equal to the reciprocal of impedance or $Y = \frac{1}{Z}$.

In ac circuits impedance also has a *phase angle* so admittance will also have a phase angle. The reader should refer to ac circuit analysis for further information on this subject.

Direct Current Circuit Analysis

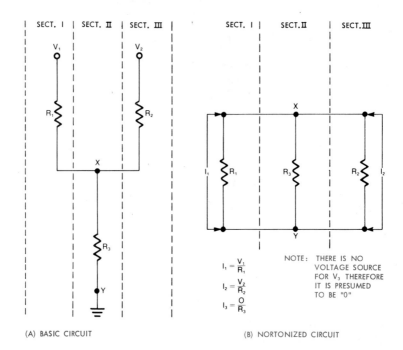

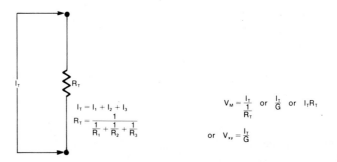

(C) EQUIVALENT NORTON CIRCUIT (D) MILLMAN'S THEOREM

Fig. 30. Millman's Theorem

Millman's theorem can be written as follows:

$$V_M = \frac{\dfrac{V_1}{R_1} + \dfrac{V_2}{R_2} + \cdots \dfrac{V_N}{R_N}}{\dfrac{1}{R_1} + \dfrac{1}{R_2} + \cdots \dfrac{1}{R_N}}$$

$$V_M = \frac{I_1 + I_2 + \cdots I_N = I_T}{\frac{1}{R_1} + \frac{1}{R_2} + \cdots \frac{1}{R_N} = \frac{1}{R_T}}$$

$$V_M = \frac{I_T}{\frac{1}{R_T}} = \frac{I_T}{G} = I_T R_T$$

Note: $\frac{V_N}{R_N}$ represents the Nth voltage source and resistance.

Millman's Theorem (Procedures)

OBJECTIVE: Find current I_{R_3} (common resistor)

① Calculate $I_1 = \frac{V_1}{R_1}$

$I_2 = \frac{V_2}{R_2}$

$I_3 = \frac{V_3}{R_3}$

$I_N = \frac{V_N}{R_N}$ (any number of sources)
(any number of resistances)

② Calculate $I_T = I_1 + I_2 + I_3 + \cdots I_N$

③ Calculate $R_T = \dfrac{1}{\frac{1}{R_1} + \frac{1}{R_2} + \frac{1}{R_3} + \cdots \frac{1}{R_N}}$ (any number of resistances)

④ Calculate G (conductance) $= \frac{1}{R_T}$

⑤ Calculate $V_{MILLMAN} = \frac{I_T}{G}$

$V_{MILLMAN} = I_T R_T$

⑥ Calculate $I_{R_3} = \frac{V_{MILLMAN}}{R_3}$

Millman's Theorem (Example)

GIVEN:
$R_1 = 3$ ohms
$R_2 = 6$ ohms
$R_3 = 4$ ohms
$V_1 = 29$ volts
$V_2 = 32$ volts

Direct Current Circuit Analysis

FIND: I_{R_3}

① Calculate $I_1 = \dfrac{29}{3} = 9{}^2/_3$ amps

$I_2 = \dfrac{32}{6} = 5{}^1/_3$ amps

② Calculate $I_T = 9{}^2/_3 + 5{}^1/_3 = 15$ amps

③ Calculate $R_T = \dfrac{1}{\dfrac{4}{12} + \dfrac{2}{12} + \dfrac{3}{12}} = \dfrac{1}{\dfrac{9}{12}} = \dfrac{12}{9} = \dfrac{4}{3}$ ohms

④ Calculate $G = \dfrac{1}{R_T} = \dfrac{3}{4} = 0.75$ mho

⑤ Calculate $V_{MILLMAN} = \dfrac{15}{0.75} = 20$ volts $\left.\begin{array}{l}V_{MILLMAN} = I_T R_T \\ = 15({}^4/_3) = 20 \text{ volts}\end{array}\right.$

⑥ Calculate $I_{R_3} = \dfrac{20}{4} = 5$ amps

Section 13

Wye and Delta Resistance Networks

Fig. 31 illustrates four configurations of three resistors each. These configurations are often used by the electrical and electronic industries in one form or another. The *pi and delta configurations* have a different form but are electrically identical. Likewise, the *tee and wye configurations* are different in form but electrically

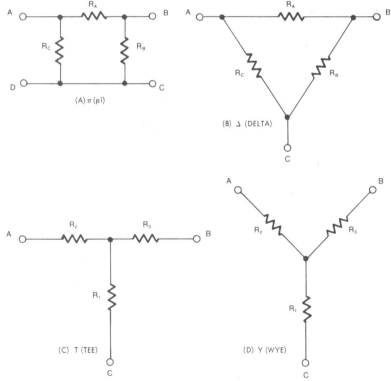

Fig. 31. Pi, Delta, Tee, and Wye Networks

Direct Current Circuit Analysis

identical. The names are derived from their shapes: *pi* for the Greek letter π, delta for the Greek letter Δ, T and Y for their English counterparts.

WYE TO DELTA RESISTANCE CONVERSION (PROCEDURES) (See Fig. 32.)

Analysis of a network may prove very difficult unless the circuit in study can be converted from wye to delta or vice-versa. In Fig. 32 a wye network is converted to a delta network. It is accomplished quite simply with the use of the following formulas:

① $$R_A = \frac{R_1 R_2 + R_2 R_3 + R_1 R_3}{R_1}$$

② $$R_B = \frac{R_1 R_2 + R_2 R_3 + R_1 R_3}{R_2}$$

③ $$R_C = \frac{R_1 R_2 + R_2 R_3 + R_1 R_3}{R_3}$$

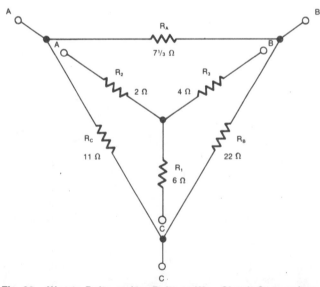

Fig. 32. Wye to Delta and/or Delta to Wye Circuit Conversions

Wye to Delta Resistance Conversion (Example)

Substituting for values shown in the circuits of Fig. 32, we have the following calculations:

① $R_A = \dfrac{(6 \times 2) + (2 \times 4) + (6 \times 4)}{6}$.

$$R_A = \dfrac{12 + 8 + 24}{6} = \dfrac{44}{6} = 7\tfrac{1}{3} \text{ ohms}$$

② $R_B = \dfrac{(6 \times 2) + (2 \times 4) + (6 \times 4)}{2}$

$$R_B = \dfrac{12 + 8 + 24}{2} = \dfrac{44}{2} = 22 \text{ ohms}$$

③ $R_C = \dfrac{(6 \times 2) + (2 \times 4) + (6 \times 4)}{4}$;

$$R_C = \dfrac{12 + 8 + 24}{4} = \dfrac{44}{4} = 11 \text{ ohms}$$

DELTA TO WYE RESISTANCE CONVERSION (PROCEDURES) (See Fig. 32.)

In Fig. 32 the values of the delta network are converted to the wye network with the use of the following formulas:

① $R_1 = \dfrac{R_B R_C}{R_A + R_B + R_C}$

② $R_2 = \dfrac{R_C R_A}{R_A + R_B + R_C}$

③ $R_3 = \dfrac{R_A R_B}{R_A + R_B + R_C}$

Delta to Wye Resistance Conversion (Example)

Substituting for values shown in Fig. 32, we have the following calculations:

① $R_1 = \dfrac{22 \times 11}{7\tfrac{1}{3} + 22 + 11}$; $R_1 = \dfrac{242}{40\tfrac{1}{3}} = 6$ ohms

Direct Current Circuit Analysis

② $R_2 = \dfrac{11 \times 7\frac{1}{3}}{7\frac{1}{3} + 22 + 11}$; $R_2 = \dfrac{80\frac{2}{3}}{40\frac{1}{3}} = 2$ ohms

③ $R_3 = \dfrac{7\frac{1}{3} \times 22}{7\frac{1}{3} + 22 + 11}$; $R_3 = \dfrac{161\frac{1}{3}}{40\frac{1}{3}} = 4$ ohms

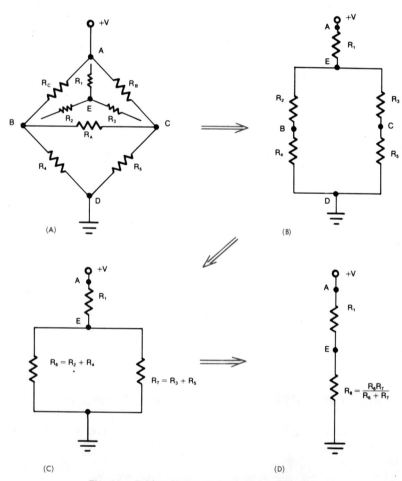

Fig. 33. Bridge Network Resistance Solution

BRIDGE NETWORK RESISTANCE SOLUTION

The bridge network solution is probably the best use made of the delta to wye network conversion. It is accomplished in several steps. These steps are shown in Figure 33.

The reader should note carefully the positions of terminals A through E in each view. In view (A), a bridge network is shown with a wye network superimposed on the top delta network. In view (B) one delta network has been converted with the use of the formulas previously mentioned to a wye network, which changes the circuit to a series parallel circuit. The terminal E represents the center of the wye network. View (C) simplifies the series parallel circuit and finally, in view (D), the series parallel circuit is further simplified to a simple series circuit.

Section 14

Nodal Method of Loop Voltage

Nodal analysis, often said to be the complement of mesh analysis, is used to solve for *node voltages* when a given circuit is excited with current sources. In nodal analysis (illustrated in Fig. 34), attention is focused on *nodes* in the circuit and the unknowns are the voltages that exist between these nodes.

METHOD OF CALCULATION USING NODAL ANALYSIS

To simplify the equations, one node is considered as reference or common. Then equations are written for the other nodes using Kirchoff's current law. The two equations are then solved using simultaneous equations, which we shall use as an example. The equations may also be solved using determinants; however, we shall leave that method to the reader.

The basic circuit is shown in Fig. 34(A). In Fig. 34(B) component and current values are added. Also, nodes are located with node 2 as the common node. In Fig. 34(C) the resistance is changed to conductance for calculation purposes. Current flow signs are added.

The reader will note the conductance G is the reciprocal of resistance, $\left(G = \dfrac{1}{R}\right)$ and is measured in mhos ($\mho$). In Fig. 34(D) the conductance values are shown along with the calculated voltage values at nodes 1 and 3, with reference to node 2. The voltage of node 1 with respect to node 3 is $V_1 - V_2$.

Nodal Circuit (Procedure)

OBJECTIVE: Find V_1 and V_2

① Simplify basic circuit. (See Fig. 34(B).)

② Assign resistance values. (See Fig. 34(B).)

③ Assign current flow signs. (See Fig. 34(C).)

Nodal Method of Loop Voltage

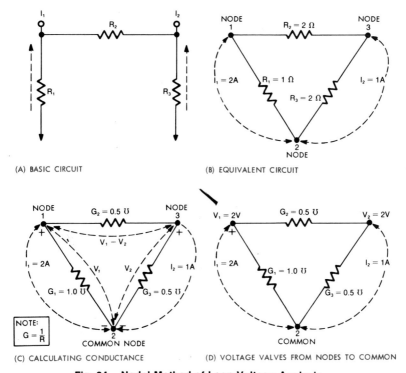

Fig. 34. Nodal Method of Loop Voltage Analysis

④ Calculate conductance values $G_1 = \dfrac{1}{R_1}$

$$G_2 = \dfrac{1}{R_2}$$

$$G_3 = \dfrac{1}{R_3}$$

⑤ Write equation for current entering node 1.
$$G_1V_1 + G_2(V_1 - V_2) = I_1$$

⑥ Write equation for current entering node 2.
$$G_3V_2 + G_2(V_2 - V_1) = I_2$$

⑦ Calculate V_1 using simultaneous equations.

⑧ Calculate V_2 by substituting in either equation.

65

Direct Current Circuit Analysis

Nodal Circuit (Example)

GIVEN:
$R_1 = 1$
$R_2 = 2$
$R_3 = 2$

$I_1 = 2A$
$I_2 = 1A$

FIND: V_1 and V_2

① Simplify basic circuit. (See Fig. 34(B).)
② Assign given resistance values as listed above.
③ Assign current flow signs.

④ $G_1 = \dfrac{1}{R_1}$

$G_1 = \dfrac{1}{1} = 1.0 \; \mho$

$G_2 = \dfrac{1}{R_2}$

$G_2 = \dfrac{1}{2} = 0.5 \; \mho$

$G_3 = \dfrac{1}{R_3}$

$G_3 = \dfrac{1}{2} = 0.5 \; \mho$

⑤ $G_1 V_1 + G_2 (V_1 - V_2) = I_1$
$1.0 \; V_1 + 0.5 (V_1 - V_2) = 2$
$1.0 \; V_1 + 0.5 \; V_1 - 0.5 \; V_2 = 2$
$\qquad 1.5 \; V_1 - 0.5 \; V_2 = 2 \quad$ (equation 1)

⑥ $G_3 V_2 + G_2 (V_2 - V_1) = I_2$
$0.5 \; V_2 + 0.5 (V_2 - V_1) = 1$
$0.5 \; V_2 + 0.5 \; V_2 - 0.5 \; V_1 = 1$
$\qquad -0.5 \; V_1 + 1.0 \; V_2 = 1 \quad$ (equation 2)

⑦ Calculate

$\qquad 1.5 \; V_1 - 0.5 \; V_2 = 2 \quad$ (equation 1)
$\qquad -0.5 \; V_1 + 1.0 \; V_2 = 1 \quad$ (equation 2)

Multiply equation 2 by 3

$\qquad 1.5 \; V_1 - 0.5 \; V_2 = 2 \quad$ (equation 1)
$\qquad -1.5 \; V_1 + 3.0 \; V_2 = 3 \quad$ (equation 2)

Algebraically combine

$$+2.5\ V_2 = 5$$
Solve $\quad V_2 = 2$ volts

(8) Substitute V_2 in equation 1.

$$1.5\ V_1 - 0.5\ V_2 = 2$$
$$1.5\ V_1 - 0.5(2) = 2$$
$$1.5\ V_1 - 1.0 = 2$$
$$1.5\ V_1 = 3$$
$$V_1 = 2\text{ volts}$$

Section 15

Reciprocity Theorem

The reciprocity theorem states that an electronic circuit is a reciprocal network if an ideal voltage source (V_S) at the input and a short circuit at the output can be interchanged between the input and output terminals without altering the current in the shorted terminals.

METHOD OF CALCULATION USING RECIPROCITY THEOREM

Fig. 35 (resistive electronic network) illustrates the voltage source, short circuit connections, and the equivalent networks. *If the networks are reciprocal, then $i_1 = i_2$ for all time.*

It should be noted that almost all invarient (passive) networks are reciprocal regardless of the network makeup. The gyrator is an exception which will not be discussed in this book because it is an isolated case.

Active circuits in general are not reciprocal networks; however, they can be in special cases. This book will only treat the passive case as shown in Fig. 35. The tee network was chosen as an example because this type circuit is recognized most readily by most electronic students. Many circuits in dc analysis can be reduced to this form.

Reciprocity Theorem (Procedure)

OBJECTIVE: Determine that current in a network is the same when voltage source at input and short circuit at output are interchanged; that is, $i_1 = i_2$. (See Fig. 35.)

Reciprocity Theorem

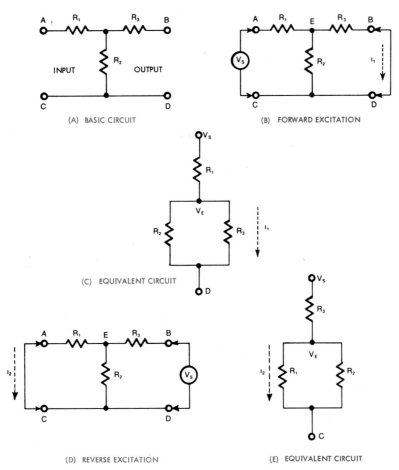

Fig. 35. Reciprocity Electronic Network

FIND: $i_1 = i_2$ (See Fig. 35.)

① Calculate $i_1 = \dfrac{V_E}{R_3}$, where $V_E = \dfrac{V_S \times \dfrac{R_2 R_3}{R_2 + R_3}}{R_1 + \dfrac{R_2 R_3}{R_2 + R_3}}$

69

$$i_1 = \frac{V_S \times \frac{R_2 R_3}{R_2 + R_3}}{R_1 + \frac{R_2 R_3}{R_2 + R_3}} \quad \text{(See Fig. 35(C).)}$$
$$R_3$$

② Calculate $i_2 = \frac{V_E}{R_1}$, where $V_E = \frac{V_S \times \frac{R_1 R_2}{R_1 + R_2}}{R_3 + \frac{R_1 R_2}{R_1 + R_2}}$

$$i_2 = \frac{V_S \times \frac{R_1 R_2}{R_1 + R_2}}{R_3 + \frac{R_1 R_2}{R_1 + R_2}} \quad \text{(See Fig. 35(E).)}$$
$$R_1$$

③ If $i_1 = i_2$ in steps ① and ② then reciprocity exists.

Reciprocity Theorem (Example)

GIVEN:
$R_1 = 10$ ohms $\qquad R_3 = 30$ ohms
$R_2 = 20$ ohms $\qquad V_S = 10$ volts

FIND: i_1 and i_2

① $\quad i_1 = \dfrac{10 \times \frac{20 \times 30}{20 + 30}}{10 + \frac{20 \times 30}{20 + 30}} = \dfrac{120}{22} = 0.182$ amps
$\qquad\qquad\qquad 30 \qquad\qquad 30$

② $\quad i_2 = \dfrac{10 \times \frac{10 \times 20}{10 + 20}}{30 + \frac{10 \times 20}{10 + 20}} = \dfrac{66.667}{36.667} = 0.182$ amps
$\qquad\qquad\qquad 10 \qquad\qquad 10$

③ $i_1 = i_2 = 0.182$ amps (Reciprocity exists.)

Section 16

Compensation Theorem

When designing practical electrical circuits it is frequently necessary to insert a current meter into the circuit to verify calculated values. Since instrumentation has finite resistance, this should be taken into account when making a measurement. Normally, the resistance of such a meter is small (50 to 200 ohms) but if not compensated for, can produce significant errors in low-resistance circuits. The compensation theorem is useful in determining such instrumentation errors and should be applied where loading is of concern.

METHOD OF CALCULATION USING COMPENSATION THEOREM (See Fig. 36.)

The compensation theorem states: If the current in a circuit is I, and if the resistance R of the branch is increased by an amount ΔR, the incremental (increase or decrease) of voltage and current in such a circuit is the voltage or current that would be produced by an opposing electromotive force equal to $I\Delta R$ inserted into the circuit in which the increase, ΔR, occurs. (See Fig. 36(C).)

In the basic circuit, Fig. 36(A), currents are calculated in the normal manner. In Fig. 36(B) the meter has been inserted in current loop I_2 and measured current is I_2'. In Fig. 36(c), ΔI_2 represents the amount of error in the measured current. The meter will indicate the calculated value for I_2 if ΔI_2 can be added to I_2' through compensation.

Compensation Theorem (Procedure)

OBJECTIVE: Compare calculated current and measured current in a circuit with compensation. (I_2 and I_2' must also be calculated.)

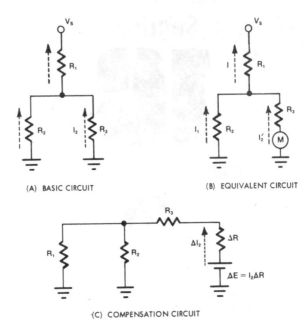

Fig. 36. **Compensation Theorem**

FIND: I_2 and I_2'

① Calculate $I = \dfrac{V_S}{R_1 + \left(\dfrac{R_2 R_3}{R_2 + R_3}\right)}$

② Calculate $I_2 = \dfrac{V_S - IR_1}{R_3}$

③ Calculate $\Delta E = I_2 \Delta R$

  where ΔR = meter resistance

④ Calculate $\Delta I_2 = -\dfrac{\Delta E}{Z + \Delta R}$

  where $Z = R_3 + \dfrac{R_1 R_2}{R_1 + R_2}$ (impedance)

⑤ Calculate $I_2' = I_2 - \Delta I_2$

        where I_2' = measured current

⑥ Calculate $I_2 = I_2' \dfrac{R'}{R' - \Delta R}$ = Compensation required to get meter to read correctly

        where $R' = Z + \Delta R$

⑦ Compare ② and ⑥ to verify compensation

Compensation Theorem (Example)

GIVEN:

$R_1 = 100$ ohms $R_3 = 200$ ohms $V_S = 10$ V
$R_2 = 200$ ohms $\Delta R = 50$ ohms (Meter resistance)

FIND: I_2 and I_2' *COMPARE:* ② and ⑥ (Calculated current and compensated current)

① $I = \dfrac{10}{100 + \dfrac{(200)(200)}{200 + 200}} = \dfrac{10}{100 + 100} = 0.05$ amps

② $I_2 = \dfrac{10 - (.05 \times 100)}{200} = \dfrac{10 - 5}{200} = 0.025$ amps

③ $\Delta E = 0.025\,(50) = 1.25$ volts

④ $\Delta I_2 = -\dfrac{1.25}{266.57 + 50} = -\dfrac{1.25}{316.67}$ $Z = 200 + \dfrac{(100)(200)}{300}$

   $\Delta I_2 = -0.00395$ amps $Z = 266.67$ ohms, impedance

Note: *The minus sign means that ΔI_2 flows in the opposite direction of I_2. Net current is positive, since I_2 is larger.*

⑤ I_2'(net current) $= 0.025 + (-0.00395)$

   $I_2' = 0.0211$ amps

⑥ $I_2 = 0.0211 \dfrac{316.67}{316.67 - 50} = 0.0211 \dfrac{316.67}{266.67}$

   $I_2 = 0.025$ amps

⑦ $0.025 = 0.025$, which verifies compensation.

Section 17

The Battery

One of the main sources of direct current is the battery. Battery action is electrical energy produced by chemical decomposition. In order that this may happen, two metal electrodes are inserted into an electrolyte. The chemical action between the two electrodes and the electrolyte causes electrical energy to build up on the terminals.

If the two electrodes are made of different materials the assembly is called a primary cell. If the two electrodes are made of the same material, the assembly is called a secondary cell. Primary cells have electrodes that are destroyed or consumed through use (example: flashlight batteries). Secondary cells have electrodes that are not destroyed in use (example: automobile lead-acid batteries).

Batteries are rated by their voltage level and their ampere-hour ratings. A battery that is rated at 10 ampere-hours will operate for 10 hours at 1 ampere. The same battery would operate 20 hours at 0.5 amperes or 5 hours at 2 amperes. Table 1 lists several different types of battery cells and the electrolyte used in them.

Other electrolytes are copper sulphate and ammonium chloride. Other electrode materials are zinc-carbon and magnesium-carbon. Some batteries can be charged, others cannot. If the battery is rechargeable, manufacturer's specifications should be checked to determine the methods, power, and time of recharging. Some batteries and their materials are potentially dangerous. Prior to

TABLE 1. BATTERY CELL ELECTRODES AND THEIR ELECTROLYTE

CELL ELECTRODES	ELECTROLYTE
ZINC-COPPER	SULPHURIC ACID
ZINC-MERCURY	POTASSIUM HYDROXIDE
NICKEL-CADMIUM	POTASSIUM HYDROXIDE
LEAD-LEAD PEROXIDE	SULPHURIC ACID
SILVER-ZINC	POTASSIUM HYDROXIDE

The Battery

use, manufacturer's special precautions should be studied to avoid damage to equipment or injury to people.

BATTERIES CONNECTED IN SERIES (See Fig. 37.)

Batteries are connected in series when voltage requirements are greater than that of the separate batteries available. Connection is from the negative pole of one battery to the positive pole of the next battery as shown in the figure. If any battery is installed in reverse, that battery will oppose the voltage of the others.

Total voltage in the series arrangement is the sum of the cell voltages. Current is the same throughout irrespective of the number of cells. Since the batteries are connected in series, the internal resistances of the batteries are also in series. Therefore series circuit analysis is applicable.

The life of each cell is measured in ampere hours. Since the same current flows through all batteries, the ampere hours for all cells is the same as each individual cell.

Formulas for Batteries in Series

$V_T = V_1 + V_2 + V_3$ (Assume each cell has a value of 1½ volts.)
$I_T = I_1 = I_2 = I_3$ (Assume each cell is rated at 2 ampere-hours.)

Series Aiding Batteries

$V_T = V_1 + V_2 + V_3$
$V_T = 1\frac{1}{2} + 1\frac{1}{2} + 1\frac{1}{2}$
$V_T = 4\frac{1}{2}$ VDC
$I_T = 2$ amp-hrs. $= 2$ amp-hrs. $= 2$ amp-hrs.

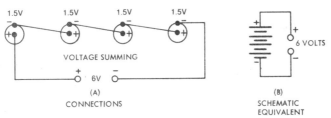

Fig. 37. Batteries Connected in Series

Series Opposing Batteries Example

$V_T = V_1 + V_2 - V_3$
$V_T = 1\frac{1}{2} + 1\frac{1}{2} - 1\frac{1}{2}$
$V_T = 1\frac{1}{2}$

BATTERIES CONNECTED IN PARALLEL (See Fig. 38.)

Batteries are connected in parallel when current requirements are greater than that of the separate batteries available. Connection is from the positive pole of one battery to the positive poles of the other batteries. Connection is also from the negative pole of one battery to the negative poles of other batteries, as shown in Fig. 38.

Total voltage in this arrangement is the same across each parallel leg. Current total is the sum of the leg currents. Since the batteries are connected in parallel, the internal resistances of the batteries are also in parallel. Therefore parallel circuit analysis is applicable.

The life of each cell is measured in ampere hours. Since the current in the parallel arrangement is additive, ampere hours are also additive. Cells of the *same value* should be placed in parallel; otherwise, the larger cell voltage will discharge first and not retain its original voltage value.

Formulas for Batteries in Parallel

$V_T = V_1 = V_2 = V_3$
$I_T = I_1 + I_2 + I_3$

Parallel Batteries Example

$I_T = I_1 + I_2 + I_3$
$I_T = 2A + 2A + 2A$ (in ampere hours)
$I_T = 6$ ampere-hours

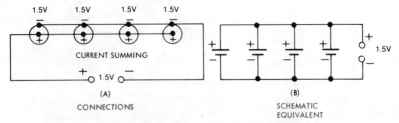

Fig. 38. Batteries Connected in Parallel

Section 18

Electrical Measurements for DC Circuits

HOW TO MEASURE CURRENT IN A DC CIRCUIT
(See Fig. 39.)

Current measurements in any circuit must be made with the ammeter placed in series with the leg to be measured. In the case of a dc circuit, the ammeter is polarized. That is, the ammeter must be placed in the circuit so that the negative and positive poles of the ammeter are placed in the negative and positive direction of current flow in the circuit. If this is not the case the ammeter needle will peg to its extremity and perhaps damage the instrument.

In Fig. 39 the ammeter is placed between points *A* and *B* in the series circuit. The ammeter could also be placed between points *C* and *D* or between points *E* and *F* without changing the ammeter indication.

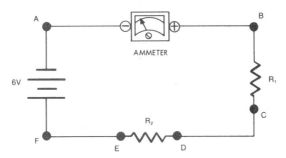

Fig. 39. Measuring Current in a DC Circuit

77

Direct Current Circuit Analysis

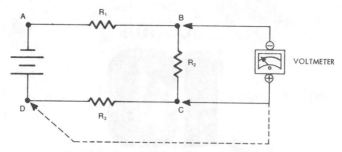

Fig. 40. Measuring Voltage in a DC Circuit

HOW TO MEASURE VOLTAGE IN A DC CIRCUIT.
(See Fig. 40.)

Measurement of voltage in a dc circuit is performed by placing the voltmeter in parallel across the resistance which is to be measured. The voltmeter measures the difference of potential between two points.

In Fig. 40 you will note that the voltmeter is placed across resistor R_2. From these points, the voltmeter measures the voltage drop across the resistor R_2. If the voltmeter leads were placed across resistors R_2 and R_3 as shown by the broken line, the voltage drop across the resistors R_2 and R_3 would be measured. That is, the difference of potential (voltage) between the two points is measured.

HOW TO MEASURE RESISTANCE.

Resistance is measured with the ohmmeter leads connected across the resistance to be measured. This places the ohmmeter in parallel with the resistance, as shown in Fig. 41.

Resistance of a resistor internal to a circuit cannot be measured efficiently because the rest of the circuit influences resistance, and accurate measurement cannot be made. If it is necessary to measure a resistance in a circuit, remove power from the circuit. Then cut or unsolder one of the resistor leads from the circuit. Then accurate resistance measurement can be made.

An ohmmeter does not require power from the circuit being measured because it uses a battery to supply its own current and voltage.

Electrical Measurements for DC Circuits

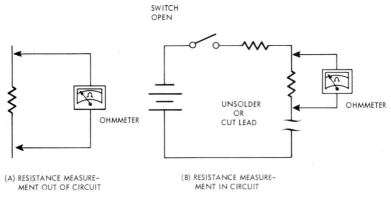

Fig. 41. Measuring Resistance

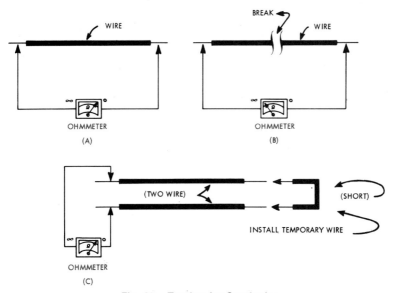

Fig. 42. Testing for Continuity

HOW TO MEASURE CONTINUITY (See Fig. 42.)

Continuity is a measure of wire resistance. Since a wire has very little resistance, placing ohmmeter leads on its extremities produces zero (0) or very low resistance, as shown in Fig. 42(A). A low scale should be selected on the ohmmeter.

In Fig. 42(B) the open in the wire causes a large resistance measurement. In Fig. 42(C), continuity is measured by placing a temporary short on two wire ends, then placing the ohmmeter leads on the other ends. This action will also show continuity by a zero (0) or very low resistance.

APPENDIX

Electrical/Electronic Safety

GENERAL

Safety education today has become an important phase of every training program. Under the 1970 Federal *Occupational Safety and Health Act* (OSHA), the employer is required to furnish a place of employment free of known hazards likely to cause death or injury. The *employer* has the specific duty of complying with safety and health standards as set forth under the 1970 act. At the same time, *employees* also have the duty to comply with these standards.

A full treatment of the subject of safety is far beyond the scope of this book, and there is ample justification for a full course in safety procedures, including first aid treatment, for all electrical and electronic technicians. Instructors should be certified by the state and qualified for any special electrical applications. The intent of this chapter is to make students fully aware of the ever-present, invisible, and generally silent hazards in handling electrical apparatus and to point out some fairly common causes of electrical shocks and fires that can easily be overlooked.

Since the time when Benjamin Franklin flew his famous kite it has become more and more apparent that electricity, even in its milder forms, is *dangerous*. A fact not widely mentioned in history is that shortly after Franklin's experiment a Russian experimenter was killed in his attempt to duplicate the kite trick. We can therefore assume that good safety habits are mandatory for all who use, direct the use of, or come in contact with electricity. Electrical equipment is found in every place that the ordinary person may find himself. For this reason it becomes the responsibility of each of us to be knowledgeable of electrical safety and to become his own and his brother's keeper.

The most basic cause of electrical accidents, as other types of accidents, is *carelessness* and the best prevention is *common sense*. However, knowing exactly what to do in an emergency is only achieved through formal education or experience. Unfortunately, the experience could be fatal, so it is more desirable to derive your knowledge through schooling. When an emergency happens it is often accompanied by panic that can cause the mind or muscles to become paralyzed. The antidote for this panic is education.

RESPONSIBILITY

Electrical/electronic safety is especially important to the technician who is exposed to electrical equipment in the raw state. He is likely to be the one who comes into direct contact with electrical and/or electronic components that may do harm to the body. Since this is the case, he finds himself in the precarious position of being continually alert to hazards that may affect him, his associates, and the people who use the equipment he builds or repairs. The supervisor is responsible for enforcing the rules of safety in the area under his direct supervision. Inspectors ensure that equipment is tested before it is released for use, and final testing should include safety precautions. Finally, the user of the equipment should be qualified in its operation so he may know when it is operating in the proper manner. *No one who comes into contact with such equipment is exempt from some degree of responsibility.*

The average layman may not believe he can get a severe shock from an electrical component disconnected from a power source. The electrical or electronic technician should know better. You know from your study of capacitance, for instance, that capacitors hold electrical charges which they later discharge. A fully charged capacitor disconnected from the power source can deliver a severe and possibly fatal shock, so you take the precaution of "shorting" it before removing it from the equipment. Also, as your brother's keeper, you should warn the user of this hazard.

ELECTRICAL SHOCK

There is a common belief that it takes a great volume of electricity to cause a fatal shock and that high voltage is the thing that provides the jolt to do the job. Imaginative stories about people being literally fried or jolted from their shoes are supplied freely by storytellers. Although there are true tales of this type, this is not generally the case. The bulk of electrical shocks come in small packages. Death from electrical shock happens most often from ordinary 60-hertz, 115 VAC house power. The effect, in general, is an instantaneous, violent type of paralysis. The human body contains a great deal of water and is normally somewhat acid and saline. For these reasons it is a fairly good conductor of electric current, which has no regard for human life or feelings.

Our brains are loaded with tiny nerve ends that provide transmitting service to the muscles. Muscles, in turn, provide motion for the body functions such as the heart and lungs. Now, assume that an electrical impulse that was not called for told your heart to speed up its pumping job or to stop its pumping job. Or suppose that a similar impulse of electricity told the lungs to quit taking in air. These situations do occur and with a comparatively small amount of current flow. Depending on a multitude of complex variables, a small current such as 10 milliamps can be very unpleasant. Currents no larger than 20 milliamps can cause muscle tightening or freezing. Currents of 30 milliamps can cause damage to brain tissue and blood vessels. And damage to brain and blood vessels can, of course, be fatal.

As you know, a decrease in resistance (according to Ohm's Law) causes current to increase. Body resistance de-

Appendix A

creases when perspiring or otherwise wet. A great number of things can cause variation in resistance. The general health of the person is probably the most important variable. A body in good condition has a much better chance of recovering from electrical shock. The muscles, by being in better tone, can recover to normal from paralyzation.

The path of current flow varies the shock. For instance, if the current involves the brain or heart it is naturally more dangerous than another path. The length of exposure can also be a factor, as well as the size or surface area of the electrical contact. Large voltages can cause spastic action, but recovery can be rapid. Also, small currents can cause muscle paralysis.

In the event of paralysis, artificial respiration or massaging must take place as quickly as possible to prevent loss of body functions and damage to the brain because of lack of oxygen. A condition known as ventricular fibrilation (uncoordinated heart beats, both fast and irregular) can occur with high currents, say 50 to 100 milliamps. This action will continue until something is done externally to restore regular heartbeat.

Death normally occurs when currents reach 250 milliamps. This does not have to occur, however, as rapid first aid can save a victim whose heart has actually stopped beating or whose lungs have stopped pumping momentarily. In many cases rapid action by knowledgeable persons can prevent body damage and save lives.

RAPID RESCUE TECHNIQUES FOR ELECTRICAL EXPOSURE

A special course in first aid will ensure the proper methods. A wrong method can be worse than no action at all. The first rule of thumb for an electrical shock victim is to remove the current path. This can be done by turning the power switch off if it is readily accessible. If not accessible, the person can be detached from the current source by using an insulator of some sort such as wood, rubber, cork, or plastic. Sometimes it is more suitable and sensible to remove the power source from the victim. Whichever is the case, isolation from the current source is by far the most important and first move that can be made.

Isolation procedures can cause problems. The rescuer may find himself in the current path. Touching a person who is paralyzed to a current source provides the current source another path through the rescuer's body. Care must be taken to prevent this from happening. After isolation, artificial respiration and/or other first aid should be applied. In all cases speed is vital. Death occurs in direct proportion to time. It is therefore obvious that a person given artificial respiration in the first three minutes has a much better chance of survival than one who is given artificial respiration after five minutes.

SNEAKY ELECTRICAL CONDUCTORS

In every electrical activity there are sneaky conductors of electricity that cause continuous problems of "shorts" and therefore electrical shock. Anywhere that you have electricity it is not just the wire, the source, or the load that provides these paths for current flow. For instance, cement may seem dry and clean but have moisture in it. In this condition the cement could be a sneaky conductor. Metal floors are, of course,

83

Direct Current Circuit Analysis

good conductors. A sweating body can cause a multitude of problems. Machines of all types in the general area will serve as conductors. Steel building posts can be conductors as well as metal roofs, steel desks, pots and pans, bicycles, automobiles, refrigerators, washing machines, and just about any other metal object you may name. These sneaky conductors can cause current draw.

Make sure, then, that you are properly grounded or that the equipment you are working with or near is grounded. Use insulated tools to prevent current from flowing where it isn't supposed to. Use floor pads and keep water and oil from floors around which you are performing electrical work. Water does not mix with electricity. Keep debris and scraps picked up to avoid similar situations. Cleanliness and alertness will help avoid or eliminate sneak circuits.

BATTERY HANDLING

Three basic safety problems are associated with the handling of batteries. These are: acid burns, fires, and explosions. Acid burns may be prevented by use of battery-handling clothes and equipment. Clothes for this purpose are mostly made of rubber and include such things as aprons, gloves, boots, and special glasses.

Proper tools are essential to perform the correct procedures safely. Proper flooring will prevent falling and spilling acids.

Fire and explosion may be caused by ignition of gases given off from the charging action. These gases, when mixed with air (oxygen), provide a highly flammable and explosive situation. Gases should not be allowed to accumulate. Ventilators should be installed in battery shops to expel the dangerous gases. Smoking in the area should be prohibited. Signs should be installed warning everyone who might enter the battery shop of the dangers that are within.

HOW TO CONTROL AN ELECTRICAL FIRE

An electrical fire is caused by current flow to some circuit that cannot withstand the current level. Also, electrical fires are caused by sneak circuits which accidently draw current, for instance a "short" to the case in a motor-driven furnace. In any event, since the cause of the problem is current flow, disconnecting the current should be the first step in eliminating the problem. Remove power from the circuit preferably by throwing a switch or by isolating the fire, using insulating material such as wood or plastic. Cut wires with wooden handled hatchets or some similar device. Prevent yourself from becoming part of the circuit. After removing the current, call the fire department, then put out the fire.

Electrical fires are best extinguished with the use of carbon dioxide (CO_2) directed toward the base of the fire. Do not use foam, as it conducts electricity.

GOOD SOLDERING HABITS

Soldering irons or soldering guns all have one thing in common: they are hot. Each is hot enough to melt solder joints. The actual temperature varies with the solder type. The speed at which the soldering iron or gun melts the solder joints is dependent on the wattage of the iron or gun and the size or complexity of the joint. In all events, the soldering device must not only be protected from the handler but also from the other circuits or equipment around it.

The soldering iron or gun should be placed in a heat-sink holder between soldering actions. Heat sinks should also be used to protect electrical/electronic circuit components. Danger of work contamination is always present as dripping or stringing of solder may occur during soldering operations. Fire hazards are always present when working with heat.

Electrical fire hazards may be prevented by ensuring that power is removed from equipment being worked on. Soldering operations should take place only after proper preparation of the work area. Clean and dry work areas, the proper wattage iron, and a well laid out soldering plan help prevent soldering accidents.

Appendix

Calculating with Powers of Ten

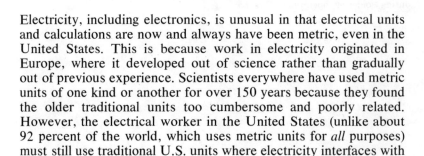

Electricity, including electronics, is unusual in that electrical units and calculations are now and always have been metric, even in the United States. This is because work in electricity originated in Europe, where it developed out of science rather than gradually out of previous experience. Scientists everywhere have used metric units of one kind or another for over 150 years because they found the older traditional units too cumbersome and poorly related. However, the electrical worker in the United States (unlike about 92 percent of the world, which uses metric units for *all* purposes) must still use traditional U.S. units where electricity interfaces with mechanical engineering, manufacturing, building trades, and other occupations and practices.

While the metric units used in electrical work are of great advantage in avoiding computation with common fractions, there is another problem to be faced and overcome. In electrical and electronic calculations the size of numbers commonly used extends over a tremendous range—from microscopic to astronomical, in fact.

ORDINARY DECIMAL NOTATION INADEQUATE

Ordinary decimal notation, quite adequate for many practical purposes, is much too cumbersome for electrical calculations. The formulas and equations used are generally stated in base units for simplicity. In practice, however, many of these base units (such as the ampere, the volt, the ohm, the farad, the henry, and the watt) are either much too large or much too small for given applications.

For instance, electric power lines may transmit several million volts (megavolts), while an electronic circuit may be sensitive to thousandths of a volt (millivolts). Similarly, while the base unit of current (ampere) is a satisfactory unit for most household appliances, thousandths of an ampere (milliamperes) or even millionths of an ampere (microamperes) are common in electronic circuits. Frequencies in hertz (cycles per second) extend from zero (in a pure dc circuit) to about 300 billion hertz (gigahertz or GHz) for weather radar and NASA communications. The base unit for capacitance is the farad, but more commonly millionths of a farad (microfarads) or even trillionths of a farad (picofarads) are more convenient for electronic work.

Consider the fairly simple equation for capacitive reactance (X_C) with frequency (f) in hertz (Hz), capacitance (C) in farads (F) and capacitive reactance (X_C) in ohms (Ω):

$$X_C = \frac{1}{2\pi fC} \quad \text{(the constant } 2\pi = 6.28 \text{ approx.)}$$

Now suppose that X_C is unknown, f = 200 MHz (that is 200,000,000 Hz), and capacitance is 372 μF (that is 0.000372 F). In the ordinary decimal notation this simple equation becomes rather a nightmare:

$$X_C = \frac{1}{6.28 \times 200,000,000 \times 0.000372}$$

$$\cong 0.00000214 \text{ ohm, or } 2.14 \ \mu\Omega \text{ (microhms)}$$

Handling numbers in this way would be extremely tedious, and errors could easily be made. Such calculations are relatively easy, however, by the "powers of ten" method.

THE POWERS OF TEN NOTATION AND METRIC PREFIXES

There is nothing at all difficult about the powers of ten notation, but it does take some practice to use it with confidence. It is simply an extension of the familiar decimal system and uses the laws of exponents applied to the base 10 to obtain the "powers of 10" that you will learn to use as multipliers. This method is indispensable for technical calculations of all kinds in either metric or customary U.S. units, but is especially adaptable to metric units. The reason is that the metric system uses prefixes that name the powers of ten used as multipliers. This is shown in Table B-1.

Positive powers of ten are known in metrics as *multiples* of the base unit and are always larger than the base unit. Thus the prefix "kilo" in kilometer means "thousand", which is 10^3 or 10 to the third power. The kilometer, a thousand times longer than the meter, is a multiple of the meter.

Negative powers of ten, always written with the minus sign before the exponent, are known as *sub-multiples* of the base unit

TABLE B-1. STANDARD PREFIXES, THEIR SYMBOLS AND MAGNITUDES

FACTOR BY WHICH UNIT IS MULTIPLIED	PREFIX	SYMBOL	OLD PREFIX & SYMBOL (IF DIFFERENT)	
10^{12}	tera-	T		
10^{9}	giga-	G	kilomega-	km
10^{6}	mega-	M		m
10^{3}	kilo-	k		
10^{2}	hecto-	h		
10^{1}	deka-	da		
10^{-1}	deci-	d		
10^{-2}	centi-	c		
10^{-3}	milli-	m		
10^{-6}	micro-	μ		
10^{-9}	nano-	n	millimicro-	mμ
10^{-12}	pico-	p	micromicro-	$\mu\mu$
10^{-15}	femto-	f		
10^{-18}	atto-	a	Kl	

EXAMPLES:

Gigahertz = GHz   Millisecond = ms

Megohm = MΩ Microhenry = μH

Kilovolt = kV   Picofarad = pF

Centimeter = cm

and are always smaller than the base unit. Thus the prefix "milli" in millimeter means "thousandth", which is 10^{-3} or 10 to the negative third power. The millimeter, a thousand times shorter than the meter, is a sub-multiple of the meter. Notice, however, that using a negative power does *not* result in a negative number.

It will be worth your while to learn the powers of ten along with their associated metric prefixes and the letter symbols for these

Appendix B

prefixes. Learn these well enough to recall them instantly. Then you are ready for the next step, which is how to express any number, large or small, in the powers of ten notation.

EXPRESSING ANY NUMBER IN THE POWERS OF TEN NOTATION

A number of any magnitude in the ordinary notation can be expressed as a single digit between 1 and 10, multiplied by a power of ten. Usually the single-digit whole number will be followed by a decimal part. For instance, 1.734 is actually a number between 1 and 2.

Taking two examples for illustration, the number 1,734,000 may be written as 1.734×10^6, and similarly the number 0.001734 may be written as 1.734×10^{-3}.

Notice that in either case the *coefficient* 1.734 is the same, because both numbers have the same digits in the same sequence. They differ only in the power of ten used: 10^6 for the large number and 10^{-3} for the small number in these examples. A point to notice is that any number which is entirely decimal in the ordinary notation, such as 0.83 or 0.0000083, is less than 1, and in the powers of ten notation will have a negative power of ten as the multiplier: in the first case $0.83 = 8.3 \times 10^{-1}$ and in the second case $0.0000083 = 8.3 \times 10^{-6}$.

In these illustrations, expressing the number as basically a single digit between 1 and 10 times the appropriate power of ten, the resulting number, such as 1.734×10^6, is in a form known as *scientific notation*. This is a very solemn name for a very simple thing. But it is not necessary to express numbers in the scientific notation form. For one reason or another, in your calculations it may be more convenient to have a different number of digits as whole numbers in the coefficient part of the expression. This is easy. 1.734×10^6 may also be written as 17.34×10^5, for instance.

To reduce the number of digits of large numbers, move the decimal point to the left as far as desired, then count the number of positions to the original decimal point. This count gives the proper exponent for the power of ten.

EXAMPLES:

(a) $470,000 = 0.47 \times 10^6$, or 4.7×10^5, or 47×10^4
(b) $31,416 = 314.16 \times 10^2$, or 3.1416×10^4, or 0.31416×10^5
(c) $285 = 2.85 \times 10^2$, or 0.285×10^3, or 0.0285×10^4

Direct Current Circuit Analysis

Often it becomes necessary to change numbers expressed in powers of ten back to the ordinary notation. An example is the approximate speed of light or of radio frequency waves, stated scientifically as 1.86×10^5 miles per second. Here the fact that no significant digits are shown beyond the third implies that this speed is known only to that degree of accuracy. In common notation this is 186,000 miles per second. This conversion is made by reversing the process previously shown: the decimal is moved back (toward the right) 5 places, as indicated by the exponent 5 in the power of ten. A little practice is recommended.

EXAMPLES:

(a) $0.47 \times 10^6 = 470{,}000$, or $4.7 \times 10^5 = 470{,}000$
(b) $314.16 \times 10^2 = 31{,}416$, or $3.1416 \times 10^4 = 31{,}416$
(c) $2.85 \times 10^2 = 285$, or $0.285 \times 10^3 = 285$

Up to this point all the numbers were larger than 1 as expressed in common notation or as positive powers of ten in the powers of ten notation. This brings us to numbers less than 1, which are the entirely decimal numbers in common notation. In the powers of ten notation these small numbers are expressed with negative powers of ten. The procedure is equally simple.

To express numbers smaller than 1 in the powers of ten notation, move the decimal point to the right as far as desired, then count the number of positions to the original decimal point. This count gives the proper negative exponent for the power of ten.

EXAMPLES:

(a) $0.000372 = 372 \times 10^{-6}$, or 37.2×10^{-5}, or 3.72×10^{-4}
(b) $0.004567 = 4.567 \times 10^{-3}$, or 45.67×10^{-4}, or 456.7×10^{-5}
(c) $0.971 = 9.71 \times 10^{-1}$, or 97.1×10^{-2}, or 971×10^{-3}

Again, it may be necessary to change numbers expressed in negative powers of ten back to the ordinary notation. This is accomplished by reversing the previous process: the decimal is moved back (toward the left) the number of positions indicated by the negative exponent of the power of ten. Zeros may have to be inserted. A little practice in this is recommended.

EXAMPLES:

(a) $372 \times 10^{-6} = 0.000372$, or $37.2 \times 10^{-5} = 0.000372$
(b) $4.567 \times 10^{-3} = 0.004567$, or $456.7 \times 10^{-5} = 0.004567$
(c) $1.293 \times 10^{-3} = 0.001293$, or $12.93 \times 10^{-4} = 0.001293$

Appendix B

One final note completes the discussion of powers of ten notation. In making calculations with powers of ten, there are occasions to subtract one exponent from another, as for example 10^{6-6}, giving 10^0. This is puzzling to many people because it doesn't appeal to common sense and seems to mean nothing. Mathematically, however, the zero power of any number, large or small, is 1. So a quantity such as $2.52 \times 10^0 = 2.52 \times 1 = 2.52$. The zero power of ten simply disappears.

USING POWERS OF TEN IN ACTUAL CALCULATIONS

The main processes of calculation which concern us are:

1. Addition
2. Subtraction
3. Multiplication
4. Division
5. Reciprocals
6. Combined Operations
7. Squares
8. Square Roots
9. Cubes
10. Cube Roots
11. Quick Approximations
12. The E Notation

Each of these processes will be explained with examples in paragraphs that follow.

1. Addition

Addition of numbers expressed in powers of ten is possible only if the same power of ten is used in all numbers. If this is not the case, first change the powers of ten to a common power. Only the significant digits are then added in the ordinary way, and the addition is multiplied by the common power of ten.

EXAMPLES:

(a) $\quad 432 \times 10^6$
$\quad\;\, + \;\, 18 \times 10^6$
$\quad\;\;\; \overline{450 \times 10^6}$

(b) $\quad 224 \times 10^{-4}$
$\quad\;\, + \;\, 32 \times 10^{-4}$
$\quad\;\;\; \overline{256 \times 10^{-4}}$

(c) $\quad 1437 \times 10^2$
$\quad\quad\;\, 13 \times 10^2$
$\quad +3415 \times 10^2$
$\quad\;\; \overline{4865 \times 10^2}$

(d) $\quad 286 \times 10^{-6}$
$\quad + \;\, 14 \times 10^2$
(Not possible to add)

2. Subtraction

Subtraction of numbers expressed in powers of ten is possible only if the same power of ten is used in both numbers. If this is not the

91

case, first change the powers of ten to a common power. Subtraction of the significant digits is then performed in the ordinary way, and the remainder is multiplied by the common power of ten.

EXAMPLES:

(a) 432×10^6
    $\underline{-18 \times 10^6}$
    414×10^6

(b) 224×10^{-4}
    $\underline{-32 \times 10^{-4}}$
    192×10^{-4}

(c) 286×10^{-6}
    $\underline{14 \times 10^2}$
    (Not possible to subtract)

3. Multiplication

A mathematical law states that when powers are multiplied their exponents are added. Powers of ten are no exception to this law of exponents. Therefore, in multiplying numbers expressed in powers of ten, the significant digits are multiplied in the ordinary way and their product is then multiplied by the new power of ten obtained by adding the exponents.

With a little practice this becomes easy.

EXAMPLE:

(a) $372{,}000 \times 3{,}200 = 3.72 \times 10^5 \times 3.2 \times 10^3$
$= 3.72 \times 3.2 \times 10^{5+3}$
$= 11.904 \times 10^8$
$\cong 11.9 \times 10^8$ (correct to 3 significant digits)

Note here that in common notation this answer would be written as 1,190,000,000—a huge, unwieldy number.

4. Division

A mathematical law states that when one power is divided by another the exponent of the divisor is subtracted from the exponent of the dividend. Powers of ten are no exception to this law of exponents. Therefore, in dividing numbers expressed in powers of ten, the significant digits are divided in the ordinary way and their quotient is then multiplied by the new power of ten obtained by subtracting the exponent of the divisor power from the exponent of the dividend.

This is actually easier to do than to explain, and with a little practice the process becomes easy.

Appendix B

EXAMPLES:

(a) $300 \div 200 = \dfrac{3 \times 10^3}{2 \times 10^2} = \dfrac{3}{2} \times 10^{3-2}$

$= 1.5 \times 10^1$

In common notation the answer is 15.

(b) $\dfrac{372,000}{1,860} = \dfrac{3.72 \times 10^5}{1.86 \times 10^3} = \dfrac{3.72}{1.86} \times 10^{5-3}$

$= 2 \times 10^2$

In common notation this would be 200.

(c) $112,000 \div 0.00028 = \dfrac{112,000}{0.00028}$

$= \dfrac{1.12 \times 10^5}{2.8 \times 10^{-4}}$

$= \dfrac{1.12}{2.8} \times 10^{5-(-4)}$

$= 0.4 \times 10^9$

$= 4 \times 10^8$

And this number in common notation is 400,000,000.

(d) $6.45 \div 0.00129 = \dfrac{6.45 \times 10^0}{1.29 \times 10^{-3}} = \dfrac{6.45}{1.29} \times 10^{0-(-3)}$

$= 5 \times 10^3$

And this number in common notation is 5,000.

Another rule of importance in division was illustrated in Problem (d) where it was found that $\dfrac{10^0}{10^{-3}} = 10^3$. *Any power of ten may be moved from the numerator to the denominator or from the denominator to the numerator if its sign is changed.*

EXAMPLES:

(a) $\dfrac{1}{10} = \dfrac{1}{10^1} = 10^{-1} = 0.1$

(Actually 10^{-1} is an alternate way of writing $1/10$.)

(b) $\dfrac{1}{100} = \dfrac{1}{10^2} = 10^{-2} = 0.01$

(c) $0.0002 = 2.0 \times 10^{-4} = \dfrac{2}{10^4}$, or in common notation $2/10000$.

Direct Current Circuit Analysis

(d) $52{,}000{,}000 = 5.2 \times 10^7 = \dfrac{5.2}{10^{-7}}$

And this number in common notation is $\dfrac{5.2}{0.0000001}$

Now as a practice exercise, start with this last number and reverse these steps.

Another rule in division of powers of ten is that like powers in both numerator and denominator may be cancelled.

EXAMPLES:

(a) $\dfrac{104 \times \cancel{10^5}}{13 \times \cancel{10^5}} = \dfrac{104}{13} = 8$

(The proof is that $10^{5-5} = 10^0$, and $10^0 = 1$)

(b) $\dfrac{344 \times \cancel{10^{-4}}}{86 \times \cancel{10^{-4}}} = \dfrac{344}{86} = 4$

(The proof is that $10^{-4-(-4)} = 10^{-4+4} = 10^0 = 1.$)

5. Reciprocals

The reciprocal of a number is defined as 1 divided by that number. Also, the reciprocal of a number written in fractional form is the number obtained by inverting the fraction. These definitions apply to numbers expressed in the powers of ten notation as well as the common notation.

Difficult reciprocal calculations are common in electrical equations, and these may be solved easily using powers of ten as shown.

EXAMPLE:

$\dfrac{1}{20{,}000 \times 0.000005 \times 0.000025}$ in common notation

$= \dfrac{1}{2 \times 10^4 \times 5 \times 10^{-6} \times 2.5 \times 10^{-5}}$

$= \dfrac{1}{2 \times 5 \times 2.5 \times 10^{4-6-5}}$

$= \dfrac{1}{25 \times 10^{-7}}$

$= \dfrac{10^7}{25} = \dfrac{10^2}{25} \times 10^5 = 4 \times 10^5$

In common notation $4 \times 10^5 = 400{,}000$.

94

Appendix B

6. Combined Operations

Many electrical equations contain combinations of large and small numbers that are to be divided and multiplied. *By first converting to the powers of ten notation and then performing the indicated calculations by rules you have already learned, you will find these can be readily solved.*

EXAMPLE:

$$\frac{0.000012 \times 8{,}000 \times 12{,}000}{3{,}000{,}000 \times 0.000004} \text{ in common notation}$$

$$= \frac{1.2 \times 10^{-5} \times 8 \times 10^{3} \times 1.2 \times 10^{4}}{3 \times 10^{6} \times 4 \times 10^{-6}}$$

$$= \frac{1.2 \times 8 \times 1.2 \times 10^{-5+3+4}}{3 \times 4 \times 10^{6-6}}$$

$$= \frac{11.52 \times 10^{2}}{12} = \frac{1152}{12} = 96$$

7. Squares

The square of a number is simply the number multiplied by itself once, and this is true of numbers expressed with powers of ten. The coefficient, or part of the number containing significant digits, is first multiplied by itself in the ordinary way, and the product is then multiplied by a new power of ten obtained by multiplying the original exponent by 2.

With a little practice this becomes easy.

EXAMPLES:

(a) $437^2 = (4.37 \times 10^2)^2$
  $= 4.37 \times 4.37 \times 10^4$
  $= 19.0969 \times 10^4$
  $\cong 19.1 \times 10^4$ (correct to 3 significant digits)
  In common notation this is 191,000.

(b) $0.00178^2 = (1.78 \times 10^{-3})^2$
  $= 1.78 \times 1.78 \times 10^{-6}$
  $= 3.1684 \times 10^{-6}$
  $\cong 3.17 \times 10^{-6}$ (correct to 3 significant digits)
  In common notation this is 0.00000317.

Direct Current Circuit Analysis

8. Square Roots

The easy way to use the powers of ten method for calculating square roots requires a little variation from the scientific notation form. *In converting from the common notation, write the coefficient, or part containing the significant digits, as a number between 1 and 100, times an even power of ten. Then take the square root of the coefficient and multiply it by a new power of ten obtained by dividing the original exponent by 2.*

With a little practice this becomes fairly easy, but a small electronic calculator or a slide rule is helpful.

EXAMPLES:

(a) $\sqrt{92800} = \sqrt{9.28 \times 10^4} = 3.0463 \times 10^{4/2}$
$\cong 3.046 \times 10^2$ (correct to 4 significant digits)
In common notation this is $304.6 \cong 305$. Notice that it is hard to estimate the magnitude of the answer when the square root of 92800 is in the common notation, but in the powers of ten notation it is obvious that the square root of 9.28 is a little over three, and this is multiplied by 10^2 or 100, so the correct answer must be a little over 300.

(b) $\sqrt{0.00348} = \sqrt{34.8 \times 10^{-4}} = 5.899 \times 10^{-4/2}$
$= 5.90 \times 10^{-2}$
In common notation this is 0.059.

9. Cubes

The cube of a number is obtained by multiplying the number by itself twice, and this is true of numbers expressed with powers of ten. The coefficient, or part of the number containing significant digits, is first multiplied twice, and the product is then multiplied by a new power of ten obtained by multiplying the original exponent by 3.

With a little practice this become easy although it is rather tedious without a calculator or slide rule.

EXAMPLES:

(a) $56.9^3 = (5.69 \times 10^1)^3 = 184.22 \times 10^3$
$= 1.842 \times 10^5$ (in scientific notation)
In common notation this is 184,200.

(b) $0.0473^3 = (4.73 \times 10^{-2})^3 = 105.82 \times 10^{-6}$
$= 1.058 \times 10^{-4}$ (in scientific notation)
This is 0.0001058 in common notation.

Appendix B

10. Cube Roots

The easy way to use the powers of ten method for calculating cube roots requires a little variation from the scientific notation form. *In converting from the common notation, write the coefficient, or part containing the significant digits, as a number between 1 and 1000 times a power of ten that has 3 as a factor, such as 10^3, 10^6, 10^9 or 10^{-3}, 10^{-6}, or 10^{-9}. Then take the cube root of the coefficient and multiply it by the new power of ten obtained by dividing the original exponent by 3.*

Manual extraction of cube roots is extremely tedious, so it is best to utilize tables or a slide rule.

EXAMPLES:

(a) $\sqrt[3]{4920} = \sqrt[3]{4.92 \times 10^3} = 1.70 \times 10^{3/3}$
$= 1.70 \times 10^1$
In common notation this is 17.
(b) $\sqrt[3]{0.000236} = \sqrt[3]{236 \times 10^{-6}} = 6.18 \times 10^{-6/3}$
$= 6.18 \times 10^{-2}$
And this is 0.0618 in common notation.

A useful fact to remember about cube roots is that the cube root of any number between 1 and 1000 is a number between 1 and 10. For instance, the cube root of 1.030301 is 1.01, and the cube root of 997.00299 is 9.99. And both 1.01 and 9.99 are between 1 and 10.

11. Quick Approximations

Many people find it very difficult to estimate the size to expect in answers to calculations when both large and small numbers occur in the same problem. The powers of ten method is ideal in cases of this kind and for making quick approximations for "ball park" figures that can later be worked out by calculator or slide rule.

To obtain a quick, rough approximation, start out by rewriting the problem entirely in standard form, in which all numbers are basically reduced to numbers between 1 and 10 times the appropriate positive or negative powers of ten. Then rewrite a second time, simplifying the coefficients, or parts containing the significant digits, writing them as the nearest whole numbers. Perform the indicated operations. Then multiply this calculation by the appropriate power of ten.

97

Direct Current Circuit Analysis

EXAMPLES:

(a) Problem as stated: $\dfrac{0.959 \times 37.5 \times 427}{185{,}000 \times 0.0637}$

In standard form: $= \dfrac{9.59 \times 10^{-1} \times 3.75 \times 10^{1} \times 4.27 \times 10^{2}}{1.85 \times 10^{5} \times 6.37 \times 10^{-2}}$

$= \dfrac{9.59 \times 3.75 \times 4.27 \times 10^{-1+1+2}}{1.85 \times 6.37 \times 10^{5-2}}$

Rough approximation:

$= \dfrac{10 \times 4 \times 4 \times 10^{2}}{2 \times 6 \times 10^{3}}$

$= \dfrac{40}{3} \times 10^{2-3}$ (after some cancellations)

$= 13.3 \times 10^{-1}$

$= 1.33$

Fully computed on electronic calculator: 1.3030664

(b) Problem as stated: $\dfrac{167{,}000 \times 29 \times 4}{30{,}000{,}000 \times 3.14 \times 8.5}$

$= \dfrac{1.67 \times 10^{5} \times 2.9 \times 10^{1} \times 4}{3 \times 10^{7} \times 3.14 \times 8.5}$

$= \dfrac{1.67 \times 2.9 \times 4 \times 10^{5+1}}{3 \times 3.14 \times 8.5 \times 10^{7}}$

$= \dfrac{1.67 \times 2.9 \times 4}{3 \times 3.14 \times 8.5} = 10^{-1}$

Rough approximation: $\dfrac{2 \times 3 \times 4}{3 \times 3 \times 9} \times 10^{-1}$

$= \dfrac{8}{27} \times 10^{-1} \cong 0.3 \times 10^{-1}$

In common notation 0.03

(Fully computed on electronic calculator: 0.0241938.) This approximation, by no means a close one, is still useful for determining the magnitude of the answer.

12. The E Notation

Still another notation is available for expressing powers of ten. This is known as the "E" notation, and is used more by metric countries than in the United States. "E" stands for exponent, and is well suited for computer processing and data transmission because it requires no writing of exponents. Here the E stands for

Appendix B

the base 10 of our decimal system, and positive or negative exponents, written with plus or minus signs follow the E.

As an example, previous Problem (b) could be written in E notation as: $\dfrac{1.67 \times 2.9 \times 4 \text{ E} + 5 + 1}{3 \times 3.14 \times 8.5 \text{ E} + 7}$

$= \dfrac{1.67 \times 2.9 \times 4 \text{ E} + 6}{3 \times 3.14 \times 8.5 \text{ E} + 7}$

$= \dfrac{1.67 \times 2.9 \times 4}{3 \times 3.14 \times 8.5} \text{ E} - 1 \cong 0.242 \text{ E} - 1,$

or 0.0242 in common notation

Using the "E" notation, this can be expressed in a single line as $(1.67 \times 2.9 \times 4) \div (3 \times 3.14 \times 8.5)$ E $- 1$. Better yet, to facilitate operations on a small electronic calculator, this can be restated as $1.67 \times 2.9 \times 4 \div 3 \div 3.14 \div 8.5 \times 0.1$. Just perform each indicated operation in sequence without stopping—first all the multiplications except the last one, then divisions, and finally the multiplication $\times 0.1$. The final multiplication stands for E -1 and converts the calculation back to common notation.

Calculators and computers produce an enormous number of digits in such cases. This is because they are utterly truthful but gullible, accepting all input data as exact. This is the fallacy of false accuracy. Using data based on instrument readings, an answer rounded to 0.0242 is more realistic than 0.0241938.

99

Appendix

Ohm's Law and Power Wheel

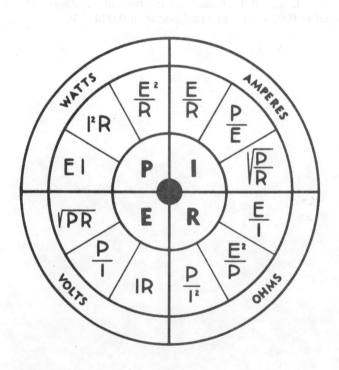

Appendix

Direct Current Formulas

SYMBOLS

E = Voltage, in volts; E_T = Total voltage
I = Current, in amperes; I_T = Total current
P = Power, in watts
R = Resistance, in ohms; R_T = Total resistance

DIRECT CURRENT

$$E = IR = \frac{P}{I} = \sqrt{PR}$$

$$I = \frac{E}{R} = \frac{P}{E} = \sqrt{\frac{P}{R}}$$

$$P = EI = I^2R = \frac{E^2}{R}$$

$$R = \frac{E}{I} = \frac{P}{I^2} = \frac{E^2}{P}$$

SERIES RESISTIVE CIRCUITS

$E_T = E_1 + E_2 + E_3 +$, etc.
$I_T = I_1 = I_2 = I_3$, etc.
$R_T = R_1 + R_2 + R_3 +$, etc.

Direct Current Circuit Analysis

PARALLEL RESISTIVE CIRCUITS

$E_T = E_1 = E_2 = E_3$, etc.
$I_T = I_1 + I_2 + I_3$, etc.

$R_T = \dfrac{R_1 \times R_2}{R_1 + R_2}$ (Two resistors only)

$R_T = \dfrac{1}{\dfrac{1}{R_1} + \dfrac{1}{R_2} + \dfrac{1}{R_3} +, \text{etc.}}$

$\dfrac{1}{R_T} = \dfrac{1}{R_1} + \dfrac{1}{R_2} + \dfrac{1}{R_3} +$, etc.

Appendix

Squares, Cubes, Roots, and Reciprocals of Numbers from 1 to 100

See pages 104 and 105 for tables.

SQUARES, CUBES, SQUARE ROOTS, CUBE ROOTS, AND RECIPROCALS OF NUMBERS FROM 1 TO 50

No.	Square	Cube	Sq. Root	Cube Root	Reciprocal
1	1	1	1.00000	1.00000	1.0000000
2	4	8	1.41421	1.25992	0.5000000
3	9	27	1.73205	1.44225	0.3333333
4	16	64	2.00000	1.58740	0.2500000
5	25	125	2.23607	1.70998	0.2000000
6	36	216	2.44949	1.81712	0.1666667
7	49	343	2.64575	1.91293	0.1428571
8	64	512	2.82843	2.00000	0.1250000
9	81	729	3.00000	2.08008	0.1111111
10	100	1,000	3.16228	2.15443	0.1000000
11	121	1,331	3.31662	2.22398	0.0909091
12	144	1,728	3.46410	2.28943	0.0833333
13	169	2,197	3.60555	2.35133	0.0769231
14	196	2,744	3.74166	2.41014	0.0714286
15	225	3,375	3.87298	2.46621	0.0666667
16	256	4,096	4.00000	2.51984	0.0625000
17	289	4,913	4.12311	2.57128	0.0588235
18	324	5,832	4.24264	2.62074	0.0555556
19	361	6,859	4.35890	2.66840	0.0526316
20	400	8,000	4.47214	2.71442	0.0500000
21	441	9,261	4.58258	2.75892	0.0476190
22	484	10,648	4.69042	2.80204	0.0454545
23	529	12,167	4.79583	2.84387	0.0434783
24	576	13,824	4.89898	2.88450	0.0416667
25	625	15,625	5.00000	2.92402	0.0400000
26	676	17,576	5.09902	2.96250	0.0384615
27	729	19,683	5.19615	3.00000	0.0370370
28	784	21,952	5.29150	3.03659	0.0357143
29	841	24,389	5.38516	3.07232	0.0344828
30	900	27,000	5.47723	3.10723	0.0333333
31	961	29,791	5.56776	3.14138	0.0322581
32	1,024	32,768	5.65685	3.17480	0.0312500
33	1,089	35,937	5.74456	3.20753	0.0303030
34	1,156	39,304	5.83095	3.23961	0.0294118
35	1,225	42,875	5.91608	3.27107	0.0285714
36	1,296	46,656	6.00000	3.30193	0.0277778
37	1,369	50,653	6.08276	3.33222	0.0270270
38	1,444	54,872	6.16441	3.36198	0.0263158
39	1,521	59,319	6.24500	3.39121	0.0256410
40	1,600	64,000	6.32456	3.41995	0.0250000
41	1,681	68,921	6.40312	3.44822	0.0243902
42	1,764	74,088	6.48074	3.47603	0.0238095
43	1,849	79,507	6.55744	3.50340	0.0232558
44	1,936	85,184	6.63325	3.53035	0.0227273
45	2,025	91,125	6.70820	3.55689	0.0222222
46	2,116	97,336	6.78233	3.58305	0.0217391
47	2,209	103,823	6.85565	3.60883	0.0212766
48	2,304	110,592	6.92820	3.63424	0.0208333
49	2,401	117,649	7.00000	3.65931	0.0204082
50	2,500	125,000	7.07107	3.68403	0.0200000
No.	Square	Cube	Sq. Root	Cube Root	Reciprocal

Appendix E

SQUARES, CUBES, SQUARE ROOTS, CUBE ROOTS, AND RECIPROCALS OF NUMBERS FROM 50 TO 100

No.	Square	Cube	Sq. Root	Cube Root	Reciprocal
51	2,601	132,651	7.14143	3.70843	0.0196078
52	2,704	140,608	7.21110	3.73251	0.0192308
53	2,809	148,877	7.28011	3.75629	0.0188679
54	2,916	157,464	7.34847	3.77976	0.0185185
55	3,025	166,375	7.41620	3.80295	0.0181818
56	3,136	175,616	7.48331	3.82586	0.0178571
57	3,249	185,193	7.54983	3.84850	0.0175439
58	3,364	195,112	7.61577	3.87088	0.0172414
59	3,481	205,379	7.68115	3.89300	0.0169492
60	3,600	216,000	7.74597	3.91487	0.0166667
61	3,721	226,981	7.81025	3.93650	0.0163934
62	3,844	238,328	7.87401	3.95789	0.0161290
63	3,969	250,047	7.93725	3.97906	0.0158730
64	4,096	262,144	8.00000	4.00000	0.0156250
65	4,225	274,625	8.06226	4.02073	0.0153846
66	4,356	287,496	8.12404	4.04124	0.0151515
67	4,489	300,763	8.18535	4.06155	0.0149254
68	4,624	314,432	8.24621	4.08166	0.0147059
69	4,761	328,509	8.30662	4.10157	0.0144928
70	4,900	343,000	8.36660	4.12129	0.0142857
71	5,041	357,911	8.42615	4.14082	0.0140845
72	5,184	373,248	8.48528	4.16017	0.0138889
73	5,329	389,017	8.54400	4.17934	0.0136986
74	5,476	405,224	8.60233	4.19834	0.0135135
75	5,625	421,875	8.66025	4.21716	0.0133333
76	5,776	438,976	8.71780	4.23582	0.0131579
77	5,929	456,533	8.77496	4.25432	0.0129870
78	6,084	474,552	8.83176	4.27266	0.0128205
79	6,241	493,039	8.88819	4.29084	0.0126582
80	6,400	512,000	8.94427	4.30887	0.0125000
81	6,561	531,441	9.00000	4.32675	0.0123457
82	6,724	551,368	9.05539	4.34448	0.0121951
83	6,889	571,787	9.11043	4.36207	0.0120482
84	7,056	592,704	9.16515	4.37952	0.0119048
85	7,225	614,125	9.21954	4.39683	0.0117647
86	7,396	636,056	9.27362	4.41400	0.0116279
87	7,569	658,503	9.32738	4.43105	0.0114943
88	7,744	681,472	9.38083	4.44797	0.0113636
89	7,921	704,969	9.43398	4.46475	0.0112360
90	8,100	729,000	9.48683	4.48140	0.0111111
91	8,281	753,571	9.53939	4.49794	0.0109890
92	8,464	778,688	9.59166	4.51436	0.0108696
93	8,649	804,357	9.64365	4.53065	0.0107527
94	8,836	830,584	9.69536	4.54684	0.0106383
95	9,025	857,375	9.74679	4.56290	0.0105263
96	9,216	884,736	9.79796	4.57886	0.0104167
97	9,409	912,673	9.84886	4.59470	0.0103093
98	9,604	941,192	9.89949	4.61044	0.0102041
99	9,801	970,299	9.94987	4.62607	0.0101010
100	10,000	1,000,000	10.00000	4.64159	0.0100000
No.	Square	Cube	Sq. Root	Cube Root	Reciprocal